T0317481

Robust Communications Software

Robust Communications Software

Extreme Availability, Reliability and Scalability for Carrier-Grade Systems

Greg Utas
Software Architect and Consultant, USA

John Wiley & Sons, Ltd

Telephone (+44) 1243 779777

Email (for orders and customer service enquiries): cs-books@wiley.co.uk
Visit our Home Page on www.wileyeurope.com or www.wiley.com

Other Wiley Editorial Offices

John Wiley & Sons Inc., 111 River Street, Hoboken, NJ 07030, USA

Jossey-Bass, 989 Market Street, San Francisco, CA 94103-1741, USA

Wiley-VCH Verlag GmbH, Boschstr. 12, D-69469 Weinheim, Germany

John Wiley & Sons Australia Ltd, 33 Park Road, Milton, Queensland 4064, Australia

John Wiley & Sons (Asia) Pte Ltd, 2 Clementi Loop #02-01, Jin Xing Distripark, Singapore 129809

John Wiley & Sons Canada Ltd, 22 Worcester Road, Etobicoke, Ontario, Canada M9W 1L1

British Library Cataloguing in Publication Data

A catalogue record for this book is available from the British Library

ISBN 0-470-85434-0

Typeset in 11/13pt Palatino by TechBooks
Printed and bound by CPI Antony Rowe, Eastbourne
This book is printed on acid-free paper responsibly manufactured from sustainable forestry in which at least two trees are planted for each one used for paper production.

To the Dragon

Contents

About the Author

Greg Utas received his honors BSc in Computer Science from the University of Western Ontario (Canada). He joined Nortel Networks in 1981, where he served as the principal software architect for various switching products. As chief software architect of GSM NSS Development, he led a team of 50 designers who redesigned the product's call processing software using object-oriented techniques. For this work, he received the Nortel Technology Award for Innovation and became the first software architect at Nortel's director level. In March 2002, Greg joined Sonim Technologies as chief software architect, responsible for the design of push-to-talk services for wireless networks. In May 2004, he left Sonim to become a consultant specializing in the design of carrier-grade software.

Greg has presented papers at the International Switching Symposium, the International Workshop on Feature Interactions in Telecommunications and Software Systems, and at ChiliPLoP, a patterns conference. He has also authored a patterns paper in *IEEE Communications* and contributed a chapter to the book *Design Patterns in Communication Systems*.

Readers can e-mail Greg at *greg@carriergradesoftware.com*. He plans to post discussions related to this book at *www.carriergradesoftware.com*.

Preface

This book is about programming techniques for building carrier-grade software systems. Carrier-grade software is required in products that support mission-critical applications. These products include routers, switches, servers, and gateways found in communication networks. Firms that operate such networks are known as carriers, so the term carrier-grade refers to a product that meets their stringent quality requirements. However, the need for carrier-grade software is now emerging in other products, such as high-end web servers, storage area network equipment, and video-on-demand servers.

Carrier-grade software must meet extreme availability, reliability, and scalability requirements. Consequently, it employs many techniques that are not common practice in the computing industry. These techniques have been proven in products designed by firms such as Lucent, Nortel, and Ericsson. However, they have never been documented in a comprehensive manner. Instead, they are lore among the software engineers who develop products that depend on them. This book attempts to elucidate what is currently a black art.

TARGET AUDIENCE

This book is primarily intended as a guide for software engineers who work on products that face carrier-grade requirements. Managers and testers of such products will also find it useful. It could also serve as a reference text for advanced software engineering courses on highly available, distributed systems that support connection-oriented protocols.

This book will also help those who perform due diligence on products that claim to be, or need to become, carrier grade. This form of due diligence often neglects to assess a product's software architecture, even though this is a primary determinant of whether a product is, or can become, carrier grade. To serve this purpose, the last chapter of this book defines maturity levels for carrier-grade software. Development teams can also use these levels for self-assessment purposes.

MOTIVATION

The impetus for this book began at the 1998 ChiliPLoP conference in Arizona. At that conference, I participated in the TelePLoP hot topic, which was mostly attended by software architects working in the telecom industry. The other group members were Ward Cunningham, Dennis DeBruler, David DeLano, Jim Doble, Bob Hanmer, John Letourneau, and Greg Stymfal. Jim Coplien organized the hot topic. Linda Rising was the conference chair and therefore could not attend our sessions, but she later edited *Design Patterns in Communications Software* [RIS01], which included patterns written by TelePLoP participants. However, few of these patterns were programming patterns, and there were many carrier-grade topics which they did not cover.

We all felt that a comprehensive book on carrier-grade software techniques was needed. Nothing close to one existed. It was as if we all lived in pre-literate societies, where knowledge was handed down in an oral tradition. We could all recite cases in which teams developing new products were unfamiliar with techniques that we took for granted. In other cases, teams exposed to these techniques dismissed them in favor of others that were common practice in the computing industry and, therefore, surely better than bizarre ideas being touted by dinosaurs. Both situations led to products that failed to meet carrier-grade requirements. These products then had to be cancelled or rewritten, or their customers had to suffer until they could replace them. We believed that a book that included the rationale for our techniques could help to avert such unfortunate outcomes.

EXAMPLES

For many reasons, most of the examples in this book come from the telecom field. It is the most mature domain when it comes to experience in constructing carrier-grade software. Conveniently, readers also have a general familiarity with the domain, given that everyone makes telephone calls. It is also an area in which I have worked as a software architect for over twenty years, so it offers me a ready source of examples that I know to be realistic rather than hypothetical.

The book contains a number of code examples. They are intended as sketches to illustrate the concepts described in the text. They are *not* carrier-grade in every detail, but they *do* convey the essence of what has been used in carrier-grade systems.

Typefaces are used as follows:

- Bold face (e.g. **five nines**) introduces technical terms.
- Small caps (e.g. OBJECT POOL) highlight names of techniques and patterns.
- Courier (e.g. `vptr`) is used for code examples and types.
- Courier italics (e.g. *`setitimer`*) denote operating system calls.

1

Introduction

This book is about programming techniques for constructing carrier-grade software systems. A **carrier** is a firm that operates a communications network which carries subscriber traffic, such as data packets or telephone calls. A **carrier-grade** system is one that meets the requirements of firms which operate these types of networks. These requirements are so strict that this book will refer to carrier-grade systems as **extreme systems**, and to carrier-grade software as **extreme software**. The reason for this is that the requirements faced by carrier-grade systems are starting to show up in other products, such as high-end web servers. Although the programming techniques described in this book have been primarily used in carrier-grade systems, other types of systems can also benefit from their use.

Extreme software is found in products which support mission critical applications. Examples of such products include the routers, switches, servers, and gateways found in communication networks where the underlying transport technology could be internet protocol (IP), asynchronous transfer mode (ATM), synchronous optical network (SONET), or a wireless transport protocol.

This chapter discusses the strict requirements that qualify carrier-grade systems as *extreme* and goes on to describe characteristics that many of these systems share. These requirements and characteristics act as forces that guide the design of the programming techniques which are the primary topic of this book.

Robust Communications Software G. Utas

1.1 REQUIREMENTS FOR EXTREME SYSTEMS

Extreme software systems face rigorous demands in the areas of **availability**, **reliability**, **scalability**, **capacity**, and **productivity**. The following sections describe the challenges presented by each of these forces.

1.1.1 Availability

Users of extreme systems want them to be available all the time: 24 hours a day, 7 days a week, every day of the year. It is therefore unacceptable to regularly shut down such systems for routine maintenance, something that is common practice in information technology (IT) networks. When a telephone switch is out of service, lives are at risk because subscribers cannot make emergency calls. High availability has been a requirement in traditional telephony networks for a long time, but is now infiltrating other networks as users become more reliant on wireless and voice over IP.

Although continuous availability is the goal of an extreme system, it cannot be achieved in practice. Cost, time to market, and the impossibility of testing all possible scenarios preclude it. Nonetheless, the requirements are stringent. The term **five nines** describes almost continuous availability. It means that a system is available 99.999% of the time, which means a maximum of about 5 minutes of downtime per year:

Nines	Downtime per Year	Downtime per Week
2 (99%)	87 hours 36 minutes	1 hour 41 minutes
3 (99.9%)	8 hours 46 seconds	10 minutes 6 seconds
4 (99.99%)	52 minutes 34 seconds	1 minute 1 second
5 (99.999%)	5 minutes 15 seconds	6 seconds
6 (99.9999%)	32 seconds	<1 second

For telephone switches, downtime even includes planned outages required for hardware maintenance and hardware and software upgrades. In the United States, the Federal Communications Commission expects telephone operating companies (telcos) to file incident reports for all outages that last more than 15 seconds. Any noticeable outage in a telephone network typically receives widespread media coverage, damaging the reputation of the telco involved. Such an outage also results in a substantial loss of revenue. These forces, when coupled with the risk of lawsuits if subscribers cannot make emergency calls, make telcos among the most demanding customers of extreme systems.

In other applications, the requirements may be less stringent. Most customers of an Internet service provider (ISP), for example, would probably be content with three- or four-nines availability. However, large corporate customers, such as on-line stockbrokers, who require more stringent service level agreements, could exclude an ISP providing less than five-nines availability from consideration. Four-nines availability might be acceptable, but only if most outages could be planned for and scheduled during non-business hours.

Achieving five-nines availability in software is difficult. Although many systems claim to be five nines, the fine print often reveals that this only refers to their hardware. In practice, software is far more likely to be the cause of outages in complex systems, which means that a five-nines hardware platform is merely a *starting point* for building an extreme system. In fact, given that the occasional software outage is all but certain, the underlying hardware platform must be *better* than five nines for the system as a whole to achieve five-nines availability. The reason is that the *product* of individual component availabilities determines overall availability when the failure of any component would cause an outage. Thus, if the software provides five-nines availability and the hardware provides six-nines availability, the system provides slightly *less* than five-nines availability ($0.999999 \times 0.99999 = 0.999989$).

Highly available systems typically replicate critical hardware components so that they can survive the loss of a replicated component. Such a loss could still cause a *partial* outage, where *some* users lose service but service continues for *most* users. In other cases, the loss of a component could cause a general service degradation, such as longer response times for all users. The replication of critical components is important because the *system* can then achieve five-nines availability even if its individual *components* do not provide five-nines availability.

1.1.2 Reliability

Users of extreme systems want them to be reliable. Whereas availability refers to a system being in service, reliability refers to it performing correctly. Even if a system is always available, users will deem it inadequate if it is full of bugs which cause it to behave in ways that are not compliant with its specifications. Reliability therefore refers to the overall quality of the system's software.

Although bug-free operation is the goal of an extreme system, it cannot be achieved in practice. Once again, cost, time to market, and the impossibility of testing all possible scenarios preclude it. The question, therefore, is how many bugs are acceptable.

In telephony systems, a typical goal is to mishandle no more than one call in 10,000, which equates to four-nines reliability, or 99.99%. In other systems, such as those that handle financial transactions, the requirements are even more stringent. If a telephone call is mishandled, the user can simply try again. However, if a financial transaction is mishandled, correcting the problem is apt to be much more tedious and expensive, because someone will probably have to correct the error manually.

The term **robustness** describes a combination of availability and reliability. A robust system remains in service and performs correctly even in the face of problems such as hardware failures, illegal arguments to functions, and software exceptions.

1.1.3 Scalability

Extreme systems must be highly scalable. A system is scalable if adding more processors allows it to support more users. The need for scalability arises for two reasons, both of which center around economies of scale:

1. The administrative cost of operating one system that caters to 100,000 users is usually far less than the cost of operating a network of ten systems, each of which caters to 10,000 users. Furthermore, such a network is likely to require closer to 15 systems, not ten, due to the overhead of interconnecting the individual systems into a network.
2. A system has a larger potential market if it is cost effective over a range of configurations. Some customers may want a large system that caters to 200,000 users, whereas other customers may want a small one that caters to 5000 users.

A large telephone switch, for example, hosts at least 100,000 users. During peak calling times, these users might generate 400,000 calls per hour. For each call, the switch receives 11 incoming events:

1. The calling user picks up the phone and receives dial tone.

2–8. The calling user dials a destination address consisting of, say, seven digits.

9. The called user answers

10. One user disconnects.
11. The other user disconnects.

The switch must therefore handle 4.4 million events per hour – over 1200 events per second. To handle this many events, the switch must be scalable: many processors must share the workload. For example, some processors simply scan the hardware that interfaces to subscriber telephones. These processors detect events such as off-hooks, on-hooks, and the dialling of digits. They then report these events to other processors which handle higher-level details, such as actually connecting calling and called users.

1.1.4 Capacity

Closely related to scalability is capacity: the throughput offered by a single processor. Even if a system is scalable, it will fail in the marketplace if it requires hundreds of processors to do what its competition can do with dozens. Its cost, either per user or per service, will be too high. Thus, a system cannot focus on scalability alone; it must also pay attention to its per-processor capacity.

Customers of extreme systems need to plan the growth of their business. When customers anticipate growth in their user base, they plan to purchase additional systems. However, they become dissatisfied if they need to purchase additional systems because of degradations in capacity, even if scalability can recover the lost capacity. Consequently, customers may insist that a new software release not degrade capacity by more than, say, 3% when compared to the previous release. This is in marked contrast to the PC marketplace, where new software releases often use so much more CPU time, memory, and disk space that users must upgrade their PCs to reattain a satisfactory level of performance.

1.1.5 Productivity

Extreme software systems are large – sometimes very large. The software for a fully featured telephone switch easily runs to millions, or even tens of millions, of lines of code. Hundreds, or even thousands, of software designers work on such systems. The product lifecycle of these systems spans decades. Most of the telephone switches in use today began their development in the 1970s or early 1980s.

Building extreme systems is therefore costly. Even if a firm develops a product that meets its customers' availability, reliability, scalability, and capacity requirements, the product's ongoing development and maintenance costs will ultimately determine its success. Designer productivity is therefore an essential element in the design of extreme software. A large number of developers must be able to work on the product in parallel, delivering new capabilities and resolving bugs in a cost-effective manner. Other things being equal, higher designer productivity increases the system's lifespan, and consequently the amount of time over which the initial investment in building it can be recouped.

It has often been observed that, inside a large software system, there is a small system struggling to get out. In other words, much of the size results from artificial, rather than fundamental, complexity. Although many extreme systems could be fairly criticized on these grounds, such criticism also belies an understanding of the forces involved. The requirements, or standards, that specify the behavior of a telephone switch are sufficiently voluminous to fill a large bookcase. A small team of developers cannot implement a system that provides this many capabilities, at least not in a timeframe that satisfies time-to-market requirements. A large development team is therefore required.

A large development team, however, is the primary cause of artificial complexity. To accommodate a large team, a system requires a solid software architecture which incorporates principles such as well-defined interfaces, layering (vertical separation), partitioning (horizontal separation), high cohesion (within components), and loose coupling (between components). The architecture must be well documented, and all software developers require training in its use. If the architecture is inadequate or poorly understood, the lack of an overall vision leads to a system with multiple, conflicting architectures. Adding new capabilities becomes difficult because of the need to fit into these conflicting architectures. Eventually the system deteriorates to the point where developers cannot predict the consequences of their changes, and so bug fixes often replace old bugs with new ones. When a system reaches this point, it must be rewritten – at a substantial cost, and at a high risk of missing its market window.

The software architecture of an extreme system must make it easy to add a new capability – such as a new hardware device, service, or protocol – without churning existing software. This reduces the risk of introducing bugs in proven code and allows far more developers to work in parallel, because they do not get in each other's way. Layering and partitioning are the primary ways of achieving these outcomes. They make it easier to add new capabilities because developers do not have to understand the entire system, only the parts in which they are working. Layering and partitioning also allow new products to deliver capabilities independently while sharing a common platform.

The software architecture should also make it possible to add a new capability without cloning and tweaking existing software. And when a new capability is not required by all customers, it should be possible to make it optional without resorting to the use of conditional compilation (`#ifdef`, for example). These regrettably common techniques, of clone-and-tweak, and the indiscriminate use of conditional compilation, lead to unmaintainable and incomprehensible code.

Because extreme systems are large and complex, they need special debugging tools. Common techniques such as setting breakpoints and writing information to the console fail to pinpoint problems quickly and cannot be used in live systems. To improve productivity and support debugging in the field, extreme systems must implement comprehensive trace tools and software logs.

1.2 BECOMING CARRIER GRADE

When a system faces extreme availability, reliability, scalability, and capacity requirements, its software must often use techniques that are not common practice in the computing industry. Some of these techniques do not necessarily apply to systems in which one or more of these requirements is absent. For example, avionics software for an aircraft flight control system must be highly reliable. It must also be highly available – but only when the aircraft is actually in use. However, it need not be highly scalable, because it only serves *one* aircraft. In contrast, the software for an air traffic control system faces all three requirements. Not only must it be reliable, it must be available whenever aircraft are flying, and it must be scalable to the point where it can track thousands of aircraft simultaneously.

The primary focus of this book is software design. This is probably the most neglected facet in the design of carrier-grade systems. Few books deal with it in a comprehensive manner. This book attempts to fill that void.

A system requires continuous improvement to become carrier grade. It does not happen immediately. Carrier gradeness is as much a journey as a destination. A good design can eventually become carrier grade. A poor design, however, will *never* become carrier grade. If you are developing a carrier-grade system, you need to be aware of the techniques described in this book. However, it is unlikely that you will use most of them in your first software release. Time-to-market pressures simply preclude this from happening. Towards the end of the book, we will therefore revisit each technique by placing it into one of these categories:

- must be implemented in your first software release;
- can be implemented in a subsequent release if your initial software design makes provisions for its eventual inclusion;
- can be implemented in a subsequent release without prior planning.

Satisfying all carrier-grade attributes – availability, reliability, scalability, capacity, and productivity – is challenging, partly because they often conflict with each other. For example, reference counting (tracking the number of references to an object so that you can delete it when all references disappear) improves reliability but degrades capacity. Application frameworks improve productivity but also degrade capacity.

A system cannot become carrier grade as the result of its software alone. All of the elements that contribute to the system must be carrier grade. The system requires carrier-grade hardware. It must be designed using a solid development process which includes rigorous stress testing. Before it is deployed, it must be modelled and engineered so that it will survive times of peak usage, even when presented with more work than it can handle. And when it is deployed, it must be easy to operate and thoroughly documented. Carrier-grade systems sometimes suffer outages as the result of procedural errors – human errors made by the craftspeople who monitor and operate these systems. When these outages result from poor documentation or overly complicated operational procedures, customers will rightly attribute them to the provider of the system rather than to the craftspeople. It is beyond the scope of this book to cover all of these topics. Fortunately, many publications discuss development processes, hardware design, system engineering (quality of service guarantees, for example), documentation, and operability.

Building a carrier-grade system is challenging and costly. You therefore need to decide, early in your product's lifecycle, whether it eventually needs to become carrier grade. If the answer is yes, you must plan its evolution accordingly and guard against being distracted from your path. Getting distracted is easy because carrier gradeness is usually important only in mature products. For a new product, content usually determines success. At first, customers primarily want you to deliver new capabilities quickly. Later, as the product and the customers' businesses mature, customers focus more on predictability. They want you to deliver software releases on time, at predictable intervals, so that they can plan the deployment of new capabilities. Carrier gradeness only becomes a customer focus later, when the product is commoditized. At this point, carrier gradeness is a differentiator, but you will have little chance to attain it unless you planned for it from the outset. Unfortunately, there have been many cases where an innovative product initially enjoyed success, only to have its lifecycle cut short when its market matured and competitors displaced it because it could not meet its customers' quality expectations.

1.3 CHARACTERISTICS OF EXTREME SYSTEMS

This section defines characteristics of extreme systems. Most extreme systems are **embedded**, **reactive**, **stateful**, **real-time**, and **distributed**. If a system does not exhibit all of these characteristics, some of the techniques that we discuss may not be the most appropriate or optimal for it.

1.3.1 Embedded

An embedded system focuses on a specific application, such as routing packets or handling telephone calls, and it often contains custom hardware developed for that application. Other applications cannot interfere with embedded applications, which is why extreme systems are usually embedded systems.

Because an embedded system is dedicated to a specific application, it runs the same software for a long time. Consequently, its designers can study its behavior before deploying it. They can build models to predict its behavior and can tune it for optimal throughput.

Because an embedded application runs alone, you can customize its operating system. Granted, this will decrease its portability, but if its platform also contains custom hardware, it is not that portable in the first place. Moreover, modifying the operating system might improve reliability or capacity. Certain characteristics of commercially available operating systems are not well suited to extreme embedded systems. We will look at these characteristics and discuss ways to modify or avoid them.

1.3.2 Reactive

A reactive system responds to external inputs. The inputs are usually messages that arrive from users or, more generally, various devices. Applications receive messages and react appropriately. When an application performs work, it often sends messages itself. Some of these are destined for users or devices, but others are destined for other applications within the system.

Each message belongs to a **protocol**, which defines

- a set of message types, which this book refers to as **signals**;
- a set of **parameters** – additional information that accompanies the signals;
- the order in which signals may be legally sent and received, and
- for each signal, whether a parameter is mandatory, optional, or illegal.

A **message** consists of a signal and zero or more parameters.

Many reactive systems lack a proper description of the protocols that drive them. If you look at their code, you can only discern the protocols by tediously studying the applications, which are themselves implemented in *ad hoc* ways. To provide traceability between their requirements and their software, reactive systems need to include protocols as key elements of their object model.

The term **transaction** refers to the work that a reactive system performs when it receives a message. When a message arrives, the system places it in a work queue. Later on, an application processes the message by performing some work, which may also involve sending some messages. When these messages are internal (that is, destined for other applications within the system), they result in additional transactions.

1.3.3 Stateful

A stateful system must remember a user's or device's current state so that it can handle subsequent events correctly. When a subscriber picks up a phone, for example, a telephone switch must know whether it is currently offering a call to the subscriber. Only in this way can it know whether to treat the off-hook event as an answer or as a request to set up a new call.

This book uses the term **session processing** to describe the behavior of stateful systems. A **session** consists of a series of transactions that are related by state. For example, a telephone call, from start to finish, comprises a single session. To be somewhat more accurate, it actually comprises two sessions: one for the calling user, and one for the called user. Each session consists of a sequence of transactions. For the calling user, the transactions are going off-hook and receiving dial tone, dialling digits and receiving ringback tone, receiving answer from the called user and connecting the speech path, and disconnecting. For the called user, the transactions are receiving a call and applying power ringing, going off-hook and connecting the speech path, and disconnecting. To handle each transaction correctly, a session must maintain state information.

In contrast to session processing, **transaction processing** refers to the handling of transactions in *stateless* systems. Systems that process financial transactions are a good example of this. They are stateless in the sense that they handle each incoming, external message (that is, messages from outside the system) atomically. A transaction may access and update a database, but it is otherwise stateless. In other words, the system does not contain sessions that are waiting for additional external inputs.

In this book, we will focus on stateful systems (session processing) rather than stateless systems (transaction processing). There are various reasons for this. First, stateless systems can be viewed as a subset of stateful systems. Second, stateful systems are generally more complex than stateless systems. Finally, many stateless systems face a different balance of forces than do stateful systems. For example, reliability is more important in financial systems than it is in communications systems, but the opposite is usually true for availability. Financial systems therefore use patterns like TWO-PHASE COMMIT [GRAY93, BERN97] to improve their reliability. This pattern, and others that apply to reliable database systems, are not covered here because they are discussed in other books and are of limited use in stateful systems.

Although our focus is on stateful systems, many of the techniques that we cover also apply to stateless systems. The reason for this is that, although a system may appear to be stateless from a black-box perspective, it is often stateful from a white-box perspective. A system that processes a financial transaction, for example, may distribute its work among many processors. An incoming, external message therefore creates a session to coordinate the sending and receiving of the internal messages that implement the transaction.

Many stateful systems lack a proper description of the state machines which underlie them. If you look at their code, you can only discern the state machines by tracing through different sequences of remote procedure calls. To provide traceability between their requirements and their software, stateful systems need to include state machines as key elements of their object model.

What a reactive, stateful system truly needs is a session processing framework which defines base classes for implementing protocols and state machines. Such a framework can dramatically improve both productivity and quality. Internally, it will embody many of the techniques described in this book. However, its overall design is a topic for another book. Until such a book becomes available, see [SGW94] and [UTAS01]. The first of these addresses the topic primarily from a modeling perspective, and the second addresses it from the perspective of a specific application domain.

1.3.4 Real-Time

A real-time system needs to perform work before a deadline occurs. There are two general types of real-time systems, based upon the strictness of their deadlines:

- A **hard real-time** system is one that *must* perform work before a deadline. If it fails to do so, it will perform incorrectly, perhaps with serious consequences.
- A **soft real-time** system is one that *should* perform work before a deadline. If it fails to do so, the consequences will be less serious, but its users will nonetheless be dissatisfied with its response time.

Most extreme systems face both hard and soft real-time requirements. Hard real-time requirements usually occur in low-level software, such as device drivers and other software which interfaces directly with hardware. Soft real-time requirements usually occur in higher-level software that provides services to end users.

Most books on real-time systems discuss techniques which apply to hard real-time systems. Some of these techniques are appropriate for soft real-time systems, but others are not. Our primary focus will be soft real-time systems, so a similar warning applies: not all of the techniques in this book are appropriate for hard real-time systems.

1.3.5 Distributed

A distributed system partitions work among many processors for a combination of the following reasons:

- To provide scalability when a single processor cannot keep up with all of the work. Here, the purpose of distribution is to offload work through delegation.
- To improve availability when another processor takes over the work of a failed processor. Here, the purpose of distribution is to improve survivability

- To simplify its design by separating software with hard real-time requirements from software with soft real-time requirements. The reason for this type of separation is discussed in Section 5.4.3.

Distribution often simplifies local (per-processor) complexity by restricting the types of work that a processor performs. At the same time, however, it increases the system's overall complexity. The first problem is how to partition the work. Another is how to handle partial failures. We will look at these issues and others in Chapter 6.

1.4 SUMMARY

- Carrier-grade systems must satisfy extreme availability, reliability, capacity, and scalability requirements. Consequently, this book often refers to them as *extreme systems*.
- Carrier-grade systems are usually embedded, reactive, stateful, real-time, and distributed in nature.
- It takes time for a system to become carrier grade. It must include certain carrier-grade techniques in its initial software release, but it will not have enough time to include all of them. It therefore needs a plan for incorporating other techniques in subsequent releases.

2

Overview

This chapter provides a high-level view of extreme software systems. It begins by defining some basic terminology. It then provides a reference model to describe the high-level architecture of a typical extreme system. Next, it defines the concept of a programming model and how such a model relates to the programming techniques that are the topic of this book. It then presents a pattern language that summarizes all of these carrier-grade techniques, and it finally concludes with an overview of the C++ classes that appear in the book.

2.1 BASIC TERMINOLOGY

This section defines some commonly used terms as they relate to this book.

System: hardware and software that is sold as a complete package and that provides an integrated set of capabilities that users value.
Network: a set of systems that interwork to offer users a broader range of capabilities or to serve a larger set of users than a single system can support.
Customer or **operator**: a firm that operates a network.
Subscriber or **user**: an end user of a network.
Service or **application**: a capability that subscribers use.
Craftspeople: customer employees who monitor and operate a network.
Engineering: configuring a system's hardware and software resources so that it provides acceptable service to subscribers when running at maximum capacity.

Robust Communications Software G. Utas
 [illegible]

Designer or **developer**: someone who writes software for an extreme system.
Node: a processing element in a system.
Shelf: a rack that houses a system's hardware.
Card: a circuit pack inserted in a shelf.
Fault: the root cause of a hardware or software error.
Error: improper behavior that arises from a fault.
Failure: an error that causes a hardware or software component to become unavailable.
Bug: a software fault that causes an error or failure.
Outage: a failure that prevents some or all subscribers from using services.

2.2 EXTREME SYSTEM REFERENCE MODEL

The high-level architecture of many extreme systems consists of the following types of nodes (see Figure 2.1):

- An **administration node** provides interfaces for craftspeople. They populate it with configuration data and subscriber profiles, which it downloads to other nodes. It receives status information from other nodes and makes it available to craftspeople.
- A **service node** runs applications on behalf of subscribers.

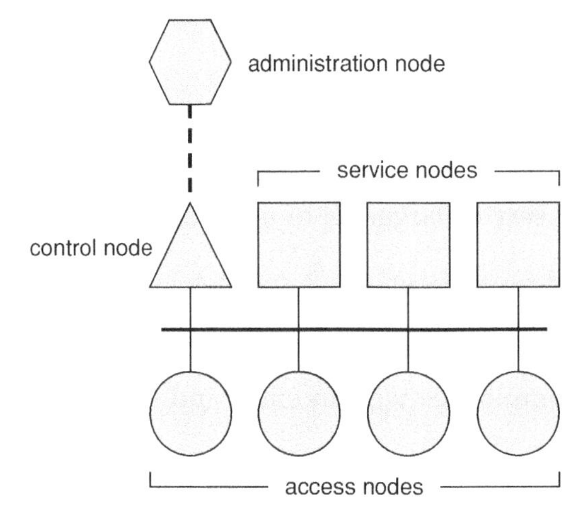

Figure 2.1 Extreme system reference model. The administration node usually resides on a separate platform and communicates with the rest of the system through the control node.

- An **access node** hosts subscriber interfaces, links to other networks, and other hardware devices.
- A **control node** runs system maintenance software that monitors the health of service and access nodes. When a node fails, the control node initiates corrective action.

The system contains one control node and a number of service and access nodes. It is managed through an administration node which sometimes acts as a front-end to more than one system. An administration node is sometimes referred to as an *element management system* if it front-ends one system, or a *network management system* if it front-ends many systems.

The control node often acts as an intermediary between other nodes. It relays administrative commands from the administration node to service and access nodes. In the other direction, it relays status information from service and access nodes to the administration node.

The administration and control nodes provide central points of control for the system. The administration node provides the external world with an integrated view of the system, and the control node contains the integrated internal view. Placing the administrative and control functions in separate nodes removes background work from service and access nodes, allowing them to focus on running services for subscribers.

Service and access nodes are often diskless. Disk accesses by their applications would unduly degrade capacity, and omitting a disk reduces costs. The data and software loads required by service and access nodes then reside on the control node's disk, and service and access nodes download it.

Although the administration node is an important part of an extreme system, it is not itself extreme. Its availability, reliability, scalability, and capacity requirements are not as strict as those of other nodes. Consequently, administration nodes usually run on general-purpose computing platforms, such as commercially available servers.

Service, access, and control nodes face extreme requirements and therefore run on hardened (high availability) computing platforms. Service and control nodes only require pure processing platforms, but hardened versions of these are available commercially. However, access nodes that support specialized hardware interfaces often run on proprietary platforms. The platforms for control, service, and access nodes are typically cards rather than standalone boxes. The cards reside in a shelf or, in a large system, in a set of interworked shelves.

2.3 PROGRAMMING MODELS

Taken as a whole, the techniques covered in this book comprise a programming model for producing carrier-grade software. A **programming model** is a set of proven techniques that all of a system's software components follow. A programming model fosters desired attributes, such as quality or simplicity, and makes it easier for components to fit together. It also provides a consistent terminology, which makes it easier for developers to communicate about their designs.

When a component deviates from the programming model, there must be a good reason for it doing so. Most often, the reason is that there was a trade-off between, say, capacity and productivity. And in the setting in question, the designer made a decision to favor one over the other.

By way of example, a programming model might contain the following elements:

- heap-based memory allocation with garbage collection (Java, for example);
- synchronous remote procedure calls (CORBA, for example);
- preemptive scheduling, using semaphores to protect critical regions;
- priority scheduling;
- spawning short-lived tasks (processes or threads) to perform work;
- frequently reading from, and writing to, disk;
- using virtual memory.

This programming model is a common one within much of the computing industry. It is also *dead wrong* for extreme systems.

What heresy!

Indeed. Yet each of the above techniques compromises carrier-grade attributes in some way. An extreme system based on the above programming model will suffer. It will exhibit less availability, reliability, capacity, scalability, and productivity than if it were based on other techniques that we will discuss.

Can an extreme system be built using the above programming model? Perhaps, but only because, as the saying goes, 'Anything can be done in software'. Some extreme systems actually do employ some of the above elements, but one can liken them to a dog walking on its hind legs. It *can* be done, but it isn't easy or elegant, and it's truly a miracle that it can be done at all.

Administration nodes, however, may freely use things that an extreme system avoids. They may do so because, as mentioned in the previous section, they do not face extreme requirements. They primarily manage databases and provide graphical user interfaces. Things like Java, CORBA, and virtual memory are often effective in such settings.

If you work, or have worked, on extreme software, you will probably be familiar with some of the techniques in this book. However, if you are primarily familiar with a programming model similar to the one outlined above, some of the techniques in this book may strike you as bizarre. Consequently, we will always discuss the rationale that underlies each of these techniques. This should help you to make trade-offs when you believe that other approaches would be more appropriate.

All the techniques in this book have been proven in carrier-grade products designed by firms such as Lucent, Nortel, and Ericsson. In many cases, they have arisen independently rather than as the result of someone bringing in techniques learned in a previous job. They have often arisen through a process of evolution, during the search for solutions to chronic software problems. However, they have never been documented in a cohesive manner. Instead, they are lore among those who develop products that depend on them. This book attempts to elucidate what is currently a black art.

2.4 A PATTERN LANGUAGE FOR CARRIER-GRADE SOFTWARE

Although this book does not use a formal pattern format to describe extreme software techniques, this section introduces these techniques as if they were patterns, using the format of a pattern language. The figure on the inside cover of this book summarizes this pattern language.

Using object orientation to build extreme systems is desirable because it allows their protocols and state machines to make use of polymorphism and inheritance. However, indiscriminate use of object orientation leads to performance problems. Well-known patterns such as SINGLETON and FLYWEIGHT mitigate this outcome, as do CACHED RESULT and EMBEDDED OBJECT. Singletons often reside in a REGISTRY that implements a POLYMORPHIC FACTORY, which preserves partitioning by allowing high-level software to delegate the instantiation of leaf classes. Defining a top-level `Object` class provides all objects with basic functions, such as `Display`, which are useful in a variety of situations.

Allocating objects from a heap at run-time leads to memory fragmentation. OBJECT POOL avoids this outcome by preallocating object blocks when a node initializes. Because this provides greater control of object management, it enables further techniques that improve performance. OBJECT TEMPLATE speeds up instantiation by initializing each object with a block-copy operation. When an object usually behaves as a singleton, its object pool uses QUASI-SINGLETON to allocate blocks efficiently. OBJECT MORPHING dynamically changes an object's class to avoid the overhead of reconstructing it. OBJECT NULLIFICATION detects references to stale objects by invalidating deleted objects.

Preemptive scheduling is undesirable for many reasons, the foremost one being that it forces the extensive use of semaphores to protect critical regions, something that is highly error prone. COOPERATIVE SCHEDULING eliminates this risk by providing a global lock that allows each transaction to run to completion. RUN-TO-COMPLETION TIMEOUT guards against transactions that run too long. A `Thread` class provides a WRAPPER FACADE and THREAD-SPECIFIC STORAGE for native threads, along with functions that support cooperative scheduling.

Under cooperative scheduling, applications avoid blocking operations. When this is impossible, a THREAD POOL is used to provide greater throughput. I/O threads receive external messages and place them in work queues that invoker threads service. These threads and work queues implement HALF-SYNC/HALF-ASYNC, which decouples applications from I/O. This allows the system to prioritize its work, which is a key requirement for the overload control techniques that are described later. When an invoker thread passes a message to a state machine, the ensuing transaction runs to completion. State machines only communicate using asynchronous messages, but this creates transient states in a group of collaborating state machines. RUN-TO-COMPLETION CLUSTER addresses this problem by providing priority intraprocessor messages so that a cluster of state machines can reach a stable state before dealing with other inputs.

In a soft real-time system, all threads perform important work, but some threads require more time than others. Priority scheduling is ill equipped to support this requirement. PROPORTIONAL SCHEDULING replaces priorities with factions. A thread's faction depends on the type of work that it performs, and all threads in the same faction receive a minimum percentage of the CPU time.

Extreme systems use distribution to increase their capacity through scalability. Distribution also improves survivability by allowing a component to be replicated. There are many ways to distribute a system's components. HETEROGENEOUS DISTRIBUTION assigns different functions to different processors. HOMOGENEOUS DISTRIBUTION assigns different users to different processors, which is often simpler and more scalable. HIERARCHICAL DISTRIBUTION combines these approaches, typically so that distribution *across* layers is heterogeneous and distribution *within* layers is homogeneous. Classic SYMMETRIC MULTI-PROCESSING is risky because it reintroduces the widespread need for semaphores. It must therefore use shared memory only for message passing. HALF-OBJECT PLUS PROTOCOL replicates part of an object in another processor to improve capacity.

Extreme systems must avoid software faults. DEFENSIVE CODING means checking pointers, array indices, arguments, and return values. It also means checking incoming messages for errors and using timeouts when sending requests. The most dangerous software faults are those which cause memory corruption. LOW-ORDER PAGE PROTECTION guards against null pointers. STACK OVERFLOW PROTECTION guards against a thread that overruns its stack. USER SPACES firewall applications to prevent them from corrupting each other's data, but this often causes significant performance penalties. WRITE-PROTECTED MEMORY provides read-only access to critical data, eliminating the penalties while still safeguarding against corruption.

When faults occur, an extreme system must detect them and initiate recovery procedures. WATCHDOG monitors a component's sanity with a timer, which the component must periodically reset. If the component fails, the timer expires and the watchdog resets the component. HEARTBEATING is a software implementation of a watchdog. LEAKY BUCKET COUNTER flags an error situation when more than a threshold number of events occur within a defined interval. This prevents the system from overreacting to intermittent faults.

SAFETY NET catches exceptions and signals to prevent a thread from being killed. It then tells the thread to clean up the work that it was performing so that it can remain in service and handle more work. An AUDIT monitors the sanity of key resources. It recovers orphaned resources and fixes corrupt data structures. If a node eventually reaches a point where it must be reinitialized, ESCALATING RESTARTS attempts to return it to service as fast as possible, with a minimal disruption of service. INITIALIZATION FRAMEWORK supports this by structuring the software that `main` invokes to bring the node into service. BINARY DATABASE supports it by reloading configuration data without having to reapply a lengthy sequence of database transactions.

An extreme system is primarily driven by messages, so its messaging system must be both efficient and robust. RELIABLE DELIVERY ensures that internal messages arrive if the destination is reachable and in service. This frees most applications from having to retransmit messages. MESSAGE ATTENUATION avoids message floods by bundling messages or sending them over a period of time.

TLV Message uses an encoding that is efficient yet flexible enough to support complex protocols with many parameters. Because it encodes parameters individually, it allows Parameter Typing to treat each parameter as a `struct` which a Parameter Template initializes. If an application writes beyond the end of a parameter, Parameter Fence detects the error. Parameter Dictionary provides fast access to a message's parameters.

Many messaging techniques reduce copying, which is often the performance bane of message-driven systems. In-Place Encapsulation reserves space at the top of a message buffer so that headers can be prepended as the message travels down the protocol stack. Stack Short-Circuiting speeds up intraprocessor messaging by bypassing the lower layers of a protocol stack. Message Cascading allows an application to send a message to multiple destinations without reconstructing it. Message Relaying allows an application to forward an incoming message to the next destination without reconstructing it. Eliminating I/O Stages reduces the number of scheduling and copy operations that occur before an incoming message finally reaches the application software that will process it.

Eliminating copy operations improves performance, but eliminating messages is even better. Prefer Push to Pull achieves this outcome by pushing data to consumers rather than having them use request–response protocols. When responding to a request, No Empty Acks eliminates bare acks (acknowledgment messages) and replaces nacks (negative acknowledgments) with asynchronous failure messages. When an initial request causes a chain of requests, Polygon Protocol eliminates intermediate acks by only sending an ack in response to the last request in the chain. Callback replaces a response message with a function call, but its use must be carefully considered because it suffers from a number of drawbacks. Shared Memory eliminates messages by placing data in a memory segment that all applications can access.

When an extreme system receives more work than it can handle, it uses overload control techniques to prevent thrashing and crashing. Finish What You Start gives priority to messages that are associated with work that is already in progress. By deferring the handling of new work until progress work is finished, it prevents the system from taking on more work than it can handle. Discard New Work throws away new work to ensure that there will be enough resources to handle progress work and to avoid wasting time on stale work that is no longer of interest to clients who have given up waiting for a response. Ignore Babbling Idiots protects the system from a flood of messages arriving from a misbehaving work source. Throttle

NEW WORK hands out credits to work sources to prevent them from sending too many requests to a server.

Extreme systems improve availability by setting aside processors to take on the work of processors that fail as the result of hardware or software faults. Failover techniques reassign work from one processor to another. LOAD SHARING runs active processors in parallel. They share the workload; if one of them fails, the others take on all of the work. COLD STANDBY leaves one or more processors inactive so that they can take over the work of processors that fail. However, this form of failover is slow when an inactive processor must first download its software, and it also drops all sessions. WARM STANDBY overcomes these drawbacks by running a pair of processors in an active-standby configuration. The active processor handles all of the work but maintains loose synchronization with the standby processor, which can then immediately take over if the active processor fails. HOT STANDBY also uses an active–standby configuration but adds custom hardware to maintain tight synchronization between the processors. Although it handles hardware failures more seamlessly, it faces the problem of synchronized insanity when a software failure occurs.

For WARM STANDBY, the primary question is how to implement synchronization. Various checkpointing techniques serve this purpose. The first three periodically send updates from the active to the standby processor, typically at the end of a transaction. APPLICATION CHECKPOINTING merely provides a message path between the active and standby processors, leaving all checkpointing details to applications. OBJECT CHECKPOINTING provides a checkpointing framework, but applications still handle many of the details. MEMORY CHECKPOINTING uses hardware to track modified pages, which are then copied from the active to the standby processor. VIRTUAL SYNCHRONY takes a rather different approach. It sends each incoming message to both processors, which therefore run in parallel. This avoids the need for checkpointing, except for when a standby processor enters service and has to catch up with the active processor. In this situation, all checkpointing techniques must use some form of bulk checkpointing to bring the standby processor up to date.

Extreme systems improve availability by installing software without disrupting service. HITLESS PATCHING installs a bug fix in a running processor by inserting new object code for a function and modifying the old function to jump to the new one. ROLLING UPGRADE installs a new software release one processor at a time, so that operators need not take an entire system or network out of service to perform a massive simultaneous upgrade. In a rolling upgrade, PRO-[illegible] ensures that processors running different software loads can still communicate. HITLESS UPGRADE uses WARM STANDBY or HOT STANDBY to install a new software release with minimal disruption. It takes a standby processor out of service and loads it with new software and configuration data. The active processor then uses bulk checkpointing to resynchronize the standby processor, after which a forced failover places the standby processor in service so that the same procedure can be repeated to upgrade the previously active processor to the new release. OBJECT REFORMATTING fixes up configuration data and checkpointed objects to conform to their modified layouts in the new software release.

An extreme system provides a number of capabilities that allow craftspeople to engineer its resources, monitor its behavior, and intervene when faults occur. CONFIGURATION PARAMETERS determine the size of resource pools and customize various behaviors. OPERATIONAL MEASUREMENTS provide usage statistics that help craftspeople determine the system's throughput and engineer the size of its resource pools. LOGS inform craftspeople of important events, some of which highlight faults that require manual intervention. ALARMS indicate that the system is in trouble and remain active until the underlying problem has been corrected. MAINTENANCE software detects faults, isolates faulty components from the rest of the system, notifies craftspeople of faults, and attempts to correct or otherwise recover from these faults. Maintenance operations include removing components from service, diagnosing them, reloading or restarting them, and initiating failovers. Maintenance operations may be initiated manually by craftspeople, or automatically by the system itself.

Extreme systems are large, so they often need to disable optional software. CONDITIONAL COMPILATION is often overused but is useful for removing lab-only debug software from production builds. SOFTWARE TARGETING defines an abstraction layer to support different platforms. The abstraction layer provides a common interface (.h file) but selects an implementation (.cpp or .c file) at compile time, based on the target platform. RUN-TIME FLAG uses a boolean to disable optional software, such as field debugging tools, at run time.

Extreme systems must provide tools which allow them to be debugged safely while in service. This rules out large core dumps, writing to the console, and breakpoint debugging. SOFTWARE ERROR LOG and SOFTWARE WARNING LOG use the log system to capture information, such as stack traces, which allow software problems to be analyzed offline. OBJECT DUMP captures the objects associated with a software error log. FLIGHT RECORDER preserves important logs for a *post mortem* analysis after an outage. Truly difficult problems often require online debugging. FUNCTION TRACER captures function calls, and MESSAGE TRACER captures internal and external messages.

TRACEPOINT DEBUGGER captures the contents of data structures at tracepoints rather than breakpoints. All of these trace tools provide triggers that enable them under specific scenarios. This avoids the overhead of capturing everything, which would also cause their buffers to overflow quickly.

As an extreme system grows to support new capabilities, its capacity decreases. Managing capacity is therefore an important activity. SET THE CAPACITY BENCHMARK suggests making your first software release efficient enough to be competitive in the marketplace, but no more. Ideally, it should contain some inefficient code, which you can easily speed up later to offset the capacity degradation caused by new capabilities. For example, many of the techniques that improve capacity for objects and messages should be deferred to subsequent releases. To predict capacity accurately, various profiling tools are required. TRANSACTION PROFILER measures the cost of individual transactions, which are then fed into a model of the system's workload to predict its capacity. FUNCTION PROFILER provides data about

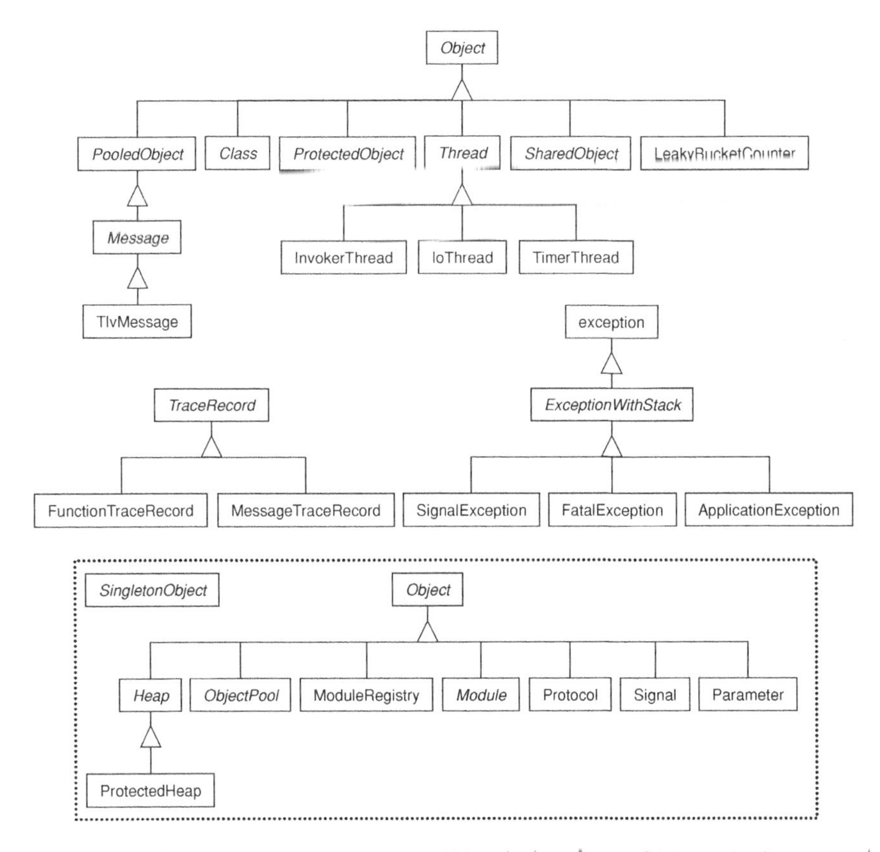

Figure 2.2 Class hierarchy. Most classes derive from Object. Instances of classes in the box are singletons, which use the SingletonObject template.

how often each function is called and how long it runs, which allows designers to focus on speeding up hotspots. THREAD PROFILER provides similar data for each thread.

2.5 CLASS HIERARCHY

To discuss design details, this book introduces various classes. Many of them receive rather cursory coverage, but a few are discussed at length. Figure 2.2 illustrates how all of these classes fit into an overall class hierarchy.

2.6 SUMMARY

- An extreme system contains an administration node, a control node, and service and access nodes.
- A programming model defines techniques that all software components use to achieve desired attributes, such as availability and reliability.
- Because a carrier-grade system must satisfy extreme availability, reliability, capacity, and scalability requirements, its programming model contains many techniques that are not commonly used in the computing industry.

3

Object Orientation

This chapter makes the case that most software in an extreme system should be written in an object-oriented (OO) programming language. It starts by describing the basic capabilities that OO provides and why these are desirable. It then presents criteria for choosing an OO language, and concludes by discussing various other topics that relate to OO.

3.1 BASIC PRINCIPLES

Many anecdotes about OO are exaggerated hype or horror stories. The crux of the matter is that OO provides three fundamental things. These things go by fancy names but form the nexus of well-architected software:

1. **Encapsulation** means that some object owns each item of data. Only the object can access its data directly. An object's class provides member functions for accessing and manipulating the object's member data. Because objects have total control over their data, they can preclude the promiscuous use of global data.
2. **Polymorphism** means that when an object's interface is similar to one that already exists, it can reuse that interface but provide a different implementation. This capability is indispensable in building application frameworks to provide high levels of abstraction (the common interfaces) while delegating work to polymorphic objects (the differing implementations). A `switch` statement whose variable is based on something like `CardType`,

Robust Communications Software G. Utas

`InterfaceType`, or `ProtocolType` is a sure sign that delegation to polymorphic objects is required.
3. **Inheritance** means that when an object is similar to one that already exists, it can reuse the existing object's software. It is so easy to do this (assuming that the existing object is reasonably well designed) that the execrable practice of clone-and-tweak all but disappears.

In Section 1.3, we characterized extreme systems as both stateful and reactive. Such systems contain many examples of both state machines and protocols, which are central to their design. To improve designer productivity and software quality, an extreme system should provide an application framework for developing state machines and protocols. Frameworks, however, depend on polymorphism and inheritance, which means that the system needs an OO language to support them. Even in the absence of a large framework, a system can benefit from OO. For example, we will eventually discuss how a wrapper class for threads helps to support extreme software requirements.

The choice of a language is important because it strongly influences design paradigms and, consequently, software architecture. It is therefore stunning how many extreme systems are still implemented in non-OO languages, primarily C. Before OO languages existed or attained sufficient maturity, extreme systems often implemented OO concepts in non-OO languages. Some systems still do so, but there is no longer a good rationale for taking this approach.

Most non-OO languages can provide encapsulation with relative ease. In practice, however, it takes a highly disciplined team to realize this attribute. Failure is likely because there are simply too many hackers out there, and some of them will be working on your product. Put them on device drivers or similar low-level components, where they can work in relative isolation.

Although a non-OO language can usually provide encapsulation without difficulty, it is harder for it to provide polymorphism, and harder still for it to provide inheritance. When polymorphism and inheritance are implemented in a non-OO language, they will be handcrafted, error-prone, inefficient versions of what is built right into an OO language. Don't waste your time on such things. An extreme system of even moderate size and complexity needs polymorphism and inheritance, which means that it needs a true OO language.

It is often reasonable to license or otherwise reuse software written in a non-OO language. Whenever possible, however, use an

OO language to implement the software that you develop from scratch.

3.2 CHOOSING A LANGUAGE

An extreme system's programming language must meet a number of criteria:

1. It should be object-oriented, for the reasons discussed in the previous section.
2. It should be type-checked to help detect programming errors.
3. It should be available on a variety of computing platforms so that you can deploy your system on multiple platforms or move to new ones when they become available.
4. It should be familiar to many developers so that you can staff your team with people who have experience with it.
5. It should perform efficiently enough to allow your system to meet its capacity requirements.
6. It should allow you to control low-level functions, such as object creation and destruction. If you cannot control these functions, it will be impossible for you to implement some of the techniques described in this book.

At present, C++ is the only language that readily satisfies all of these criteria. Code examples in this book therefore use C++. Other OO languages – Java, Eiffel, Smalltalk, and Ada – fall short in one or more areas. These languages, or others, may eventually become viable alternatives to C++. But for now, C++ is the most obvious choice.

With regard to Java, the primary issues are efficiency and control over low-level functions. To use Java for extreme software, you will probably need to do the following:

- Use compiled Java, whether in the form of pre-compilation or just-in-time compilation. This will address efficiency concerns.
- Use real-time Java [RTJ01]. This will give you control over some low-level functions which are needed to support some of the techniques discussed in this book.
- Use the Java Native Interface [LIA99] to implement the low-level functions that real-time Java does not provide.

Some extreme systems have indeed been implemented in Java. If your team has a marked preference for Java over C++, Java might be a reasonable choice.

3.3 OTHER CONSIDERATIONS

When using object orientation, it is important to focus on the fundamentals:

- encapsulation: hidden data;
- polymorphism: shared interfaces;
- inheritance: shared implementations.

However, there are also some other things to consider:

- *Operator overloading* may seem mysterious, but an operator is really just a member function that is defined and invoked using a different syntax. There are situations where operator overloading is useful or even indispensable. The primary guideline is to use an overloaded operator consistently. If an operator performs a useful function in a base class, then derived classes should only extend its behavior, not redefine it for other purposes. If an operator does not perform a useful function in a base class, you can redefine it for another purpose if you use it consistently. A good example is the way that I/O streams redefine the operators << and >> for use by `cout` and `cin`.
- *Multiple inheritance* must be used carefully, if at all. Most OO languages do without it because aggregation is often a better alternative. However, C++ does provide multiple inheritance. As a general guideline, it should only be used to implement the equivalent of Java interfaces. In C++, this is done by defining a class that declares only pure virtual functions. In other words, the class only defines an interface, but no implementation. Its interface is then mixed into primary classes using multiple inheritance. The purpose of the interface is to define auxiliary functions that a primary class must implement when it plays a role that is orthogonal to its core purpose.
- *Templates* are useful, particularly for implementing singletons. However, they should be used sparingly if bloat in the object code is a concern. And when speed is important, templates are less useful. A template that is generic enough to satisfy all possible users may not offer adequate performance. Its capabilities must then be manually merged and optimized into classes that require speed.
- *Exceptions* cause considerable overhead, both in the size of the object code and in the CPU time required to handle them. They should therefore be used very sparingly, for errors that are truly unexpected. However, for handling errors that would otherwise

be fatal, exceptions are invaluable because the `Enter` function of a base `Thread` class can catch them. This function can then initiate recovery procedures without causing a crash and without even needing to recreate the thread. We will look at how to do this in Section 8.1.

Another consideration is that some platforms, typically ones that support smaller embedded systems, only provide compilers for Embedded C++ [ECPP99]. This is a subset of C++ that does not support multiple inheritance, templates, exceptions, RTTI, or namespaces. Consequently, if you plan to run your software on this type of platform, you will have to avoid all of these things. Mobile phone platforms, for example, often impose these restrictions. Although a mobile phone is not an extreme system in the sense of this book, it might nonetheless share software with an extreme system. If, for example, you are developing a SIP or RTP protocol stack for use in both a mobile phone (a client) and a server, you may be forced into this type of least common denominator situation.

3.4 DISTRACTIONS AND MYTHS

When you adopt object orientation, a number of other topics are likely to arise. They merit some consideration, but it is easy for them to become distractions:

- *Methodologies*, including *notations* and *tool suites*, can be a considerable source of distraction. Here, it is important to be eclectic. For example, UML diagrams [RJB98, DOUG00, GAR02] are very useful for documenting frameworks. But individual designs based on these frameworks need not indulge in obvious rehashings of those diagrams. State machines, for example, are better documented by SDL diagrams [ITU99], which focus on detailed behavior.
- *Training* in OO analysis and design is helpful, but the need for such training is often overstated. Most computer science curricula have included OO since the 1980s. Moreover, it is sufficient to have a few designers who are skilled in OO techniques, for they will be able to shepherd a large team. If these designers also provide application frameworks, they will considerably simplify the design of applications based on those frameworks. Frameworks define a system's fundamental object models, thereby providing a detailed roadmap for application developers.
- The *capacity penalty* caused by OO is invariably overstated. Nevertheless, this area requires attention. Many projects have indeed

failed as the result of neglecting capacity. In the next chapter, we will look at techniques that can speed up OO software in situations where this is important. However, these techniques are rather tactical. From a strategic perspective, high-level designs, and frameworks in particular, must be carefully assessed from the standpoint of capacity. Classes should be merged until those that remain serve orthogonal purposes. This will reduce the number of objects created and destroyed at run time. Using a hundred small objects when a dozen larger ones would suffice practically guarantees unacceptable capacity. If you avoid this type of over-design, your C++ software will run almost as fast as a C version of it. Assembly code can be faster than C, but this is rarely an argument for writing a system in assembly code. The same holds true when comparing C to C++.

Even *design patterns* can be a distraction. At the 1999 ChiliPLoP conference, Jim Coplien gave a talk whose thesis was that people should document pattern languages rather than isolated patterns. He also expressed concern about the use of *Design Patterns* [GHJV95] in introductory courses on object orientation. Coplien sees patterns as tactical exceptions to encapsulation, polymorphism, and inheritance, and believes that a premature focus on patterns diverts attention from these fundamentals.

Individual patterns have one advantage over pattern languages: namely, they are (or should be) fairly domain independent, whereas pattern languages are usually domain specific. But when I consider Coplien's comments in light of *A Pattern Language of Call Processing* [UTAS01], I think his observations are insightful. This pattern language was a chapter that I contributed to a book, one that grew out of a paper that I originally presented at the 1998 International Workshop on Feature Interactions in Telecommunications and Software Systems. Nowhere in that chapter or paper, however, did I describe the role of inheritance or polymorphism in the session processing framework that embodies the pattern language. Yet a primary characteristic of that framework is the breadth and depth of the class hierarchies that arise through protocol-based subclassing.

Even a pattern language, therefore, must discuss inheritance and polymorphism. As isolated patterns, they are ubiquitous. But their role in a pattern language should be clearly stated. In retrospect, it is not surprising that the first question from the audience, after I presented the earlier version of [UTAS01] at that conference, was 'How did you use inheritance?' I am still embarrassed that this question had to be asked.

3.5 SUMMARY

- The fundamentals of OO are encapsulation, polymorphism, and inheritance.
- Frameworks improve designer productivity. An extreme system contains groups of similar applications that can leverage application frameworks.
- Frameworks require polymorphism and inheritance for their implementation, so an extreme system needs an OO language.
- C++ is a logical choice for an OO language, but compiled Java is also a candidate if you use its real-time extensions and implement some low-level capabilities using the Java Native Interface.
- Use operator overloading when necessary, but use it consistently.
- Use multiple inheritance – if at all – to inherit interfaces, not implementations.
- Use templates judiciously when code bloat or speed is a concern.
- Use exceptions only for serious errors that you do not expect application software (leaf classes) to handle.
- Some small platforms use Embedded C++, which does not support multiple inheritance, templates, or exceptions.
- Assess OO designs and software to ensure that they will not unduly degrade capacity.

4

Using Objects Effectively

This chapter discusses techniques that help an extreme system to use object orientation successfully. Although some of these techniques improve availability or productivity, most of them improve capacity.

The inefficient use of objects forces applications to circumvent good design principles in order to attain acceptable capacity. This subverts, rather than preserves, design integrity. Capacity must therefore be considered during a system's high-level design phase. Significant reengineering will be required if, for example, a system contains frameworks which incur too much processing overhead or use many small objects that could otherwise be combined into larger ones.

The cost of constructing and destroying objects is far and away the greatest overhead in OO languages. Various chapters in [MEY92] and [MEY96] discuss this topic and other efficiency considerations. In particular, avoid declaring objects on the stack and passing objects by value, both of which can incur unintended overheads.

In this section, the techniques that improve capacity are

- SINGLETON,
- FLYWEIGHT,
- CACHED RESULT,
- OBJECT POOL,
- OBJECT TEMPLATE,
- QUASI-SINGLETON,
- EMBEDDED OBJECT, and
- OBJECT MORPHING.

With the exception of CACHED RESULT, all of these techniques reduce the cost of constructing and destroying objects. SINGLETON and

[illegible]

FLYWEIGHT improve capacity dramatically by creating objects at initialization time, thus completely avoiding the cost of creating them at run time.

Two techniques in this section improve availability. They are

- OBJECT POOL and
- OBJECT NULLIFICATION.

Two other techniques for improving availability, the `ProtectedObject` class and OBJECT PATCHING, appear in later sections, after discussions of memory protection and software patching, respectively.

The remaining techniques in this section improve productivity by focusing on software architecture:

- `Object` class,
- REGISTRY, and
- POLYMORPHIC FACTORY.

4.1 THE `Object` CLASS

Before we define any domain-specific classes, we will start by defining an abstract class whose functions are useful for all objects. For example, each class should provide a function which displays an object in text format, a capability that is useful for debugging and logging purposes. This capability must be ubiquitous to support an OBJECT BROWSER, which allows designers to view objects in a running system. We therefore define it in the class `Object`, which resides at the top of the class hierarchy:

```
typedef void* ClassVptr;

class Class; // forward declaration; will be discussed
             // later

class Object
{
public:
   virtual ~Object(void);
   static string Spaces(int count);
   static string StrObj(const Object *obj,
                        bool thisptr);
   string ClassName(void) const;
```

```
   virtual ClassVptr GetVptr(void);
   virtual void Display(ostream &stream, int indent,
                        bool verbose);
   virtual void GetSubtended(Object *objects[],
                             int &count);
   virtual void Patch(int selector, void *arguments);
protected:
   Object(void);
   void MorphTo(Class &target);
private:
   Object *patchData_;
};
```

Every class with member data implements `Display` by invoking `Display` on its superclass and then displaying its own member data. Although overriding `operator <<` is one way to support a display capability, `Object` defines a `Display` function to support arguments more easily:

- `stream` specifies where to display the object. It often references a file or an `ostringstream` rather than `cout`.
- `indent` specifies how much to indent the output. This illustrates ownership when displaying a group of objects, making the output easier to read. If *A* owns *B*, whether through composition (*B* is a member of *A*) or association (*A* has a pointer to *B*), *A*'s `Display` function increases the indentation when it invokes *B*'s `Display` function. `Object::Spaces` supports indentation by returning a string of `count` blanks:

```
string Object::Spaces(int count)
{
   string s = "";
   if(count > 0) s.insert(s.begin(), count, ' ');
   return s;
}
```

- `verbose` controls the amount of output. If it is set, everything is displayed, else the output is abbreviated. For example, if `SomeObject` owns two objects, its `Display` function would look like this:

```
void SomeObject::Display(ostream &stream, int indent,
                         bool verbose)
{
   // must invoke superclass first
   Object::Display(stream, indent, verbose);
   indent = indent + 3;
```

```
      stream << Spaces(indent) << "a : ";
      if((a_ != NULL) && (verbose))
      {
         stream << endl;
         a_->Display(stream, indent + 2, verbose);
      }
      else stream << StrObj(a_, true) << endl;
      stream << Spaces(indent) << "b : ";
      if((b_ != NULL) && (verbose))
      {
         stream << endl;
         b_->Display(stream, indent + 2, verbose);
      }
      else stream << StrObj(b_, true) << endl;
   }
```

`Object::StrObj` returns a string that contains an object's class name and, optionally, a pointer to the object's location:

```
string Object::StrObj(const Object *obj, bool thisptr)
{
   if(obj != NULL)
   {
      ostringstream stream;
      // display pointer to object
      if(thisptr) stream << obj << " ";
      stream << obj->ClassName();
      return stream.str();
   }
   return "NULL";
}
```

`Object::ClassName` cleans up an object's class name. Its implementation is compiler specific. The string returned by `typeid` is unlikely to be the actual class name found in the source code. Internally, a compiler may add a prefix or suffix, perhaps a digit string or the string `'class'`. If you don't want this noise in your output, you have to remove it:

```
string Object::ClassName(void) const
{
   string name = typeid(*this).name();
   if(name.compare(0, 6, "class ") == 0)
      return name.substr(6, name.size());
   return name;
}
```

`Object::Display` uses the `ClassName` function itself:

```
void Object::Display(ostream &stream, int indent,
                     bool verbose)
{
   stream << Spaces(indent) << ClassName() << endl;
   indent = indent + 2;
   stream << Spaces(indent) << "this      : ";
   stream << this << endl;
   stream << Spaces(indent) << "patchData : ";
   if((patchData_ != NULL) && (verbose))
   {
      stream << endl;
      patchData_->Display(stream, indent + 2,
                          verbose);
   }
   else stream << StrObj(patchData_, true) << endl;
}
```

`Object::GetSubtended` generates a list of all the objects that are transitively owned by a given object. This function is useful for displaying or cleaning up an entire tree of objects. At the `Object` level, its implementation is

```
void Object::GetSubtended(Object *objects[],
                          int &count)
{
   objects[count] = this;
   count++;
   if(patchData_ != NULL)
      patchData_->GetSubtended (objects, count);
}
```

An object that owns other objects overrides `GetSubtended`:

```
void SomeObject::GetSubtended(Object *objects[],
                              int &count)
{
   // must invoke superclass first
   Object::GetSubtended(objects, count);
   if(a_ != NULL) a_->GetSubtended(objects, count);
   if(b_ != NULL) b_->GetSubtended(objects, count);
}
```

Invoking `GetSubtended` on object *A* causes a depth-first traversal of ownership relationships in the underlying object model to build a list of all the objects that *A* transitively owns.

For now, we will leave the `Object` class. Later on, we will discuss its other member functions and derive other classes from it.

4.2 BASIC DESIGN PATTERNS

This section summarizes some fundamental design patterns that are often useful for implementing extreme software. Some of these patterns appear in [GHJV95], so they only receive cursory coverage here.

4.2.1 Singleton

A SINGLETON [GHVJ95] is an object of which there is only one instance in a running system. It sometimes serves to wrap global data or to control access to a global resource, in which case its primary purpose is encapsulation.

A singleton should usually be created during system initialization and remain permanently allocated. It is then immediately available once the system is in service. This avoids any unexpected delays that could occur if the singleton had to be constructed while performing payload work (that is, the primary work performed by the system).

A template class readily supports singletons:

```
template<class T> class SingletonObject
{
public:
   inline SingletonObject(void)   { Instance(); };
   inline ~SingletonObject(void)  { FreeInstance(); };

   inline static T *Instance(void)
   {
      if(instance_ == NULL) instance_ = new T();
      return instance_;
   };

   inline void FreeInstance(void)
   {
      delete instance_;
      instance_ = NULL;
   };
private:
   static T *instance_; // the singleton instance
};

// For initializing each singleton's instance_
//
template<class T> T *SingletonObject<T>::instance_ =
          [illegible]
```

The singleton for `SomeSingleton` is then accessed by

```
SomeSingleton *s =
   SingletonObject<SomeSingleton>::Instance();
```

If the singleton does not yet exist, this statement has the side effect of creating it. Each class whose object will be a singleton must define its constructor and destructor as `protected`, so that it can only be instantiated through its singleton template. It must also make the singleton template a friend class, in order to give it access to the protected constructor and destructor:

```
class SomeSingleton: public Object
{
   friend class SingletonObject<SomeSingleton>;
protected:
   // template precludes arguments to constructor
   SomeSingleton(void);
   ~SomeSingleton(void);
};
```

The constructor cannot have any arguments because the template's `Instance` method has no way to support them. If `SomeSingleton` should not be subclassed, its constructor and destructor should be `private` rather than `protected`.

4.2.2 Flyweight

A FLYWEIGHT [GHJV95] is an object, often a SINGLETON, that many would-be instances of a class share. Flyweights improve capacity when it is too expensive, in terms of processor or memory usage, to actually instantiate a separate object for each would-be instance of the class.

Because a flyweight is shared, its member data is limited to things that apply to all of its would-be instances. However, some of a flyweight's functions may provide an argument for passing what is effectively member data. This member data is temporary, however, because it can only be used within the function in question. Consequently, flyweights are stateless.

4.2.3 Registry

A REGISTRY accesses an object using an identifier. The identifier is often an `int`, in which case an array, indexed by the identifier, can implement the registry. Each array element is a pointer to a registered object. The registered objects are often FLYWEIGHT SINGLETONS, which add themselves to the registry during system initialization.

When a registry contains flyweight singletons, the identifier that indexes the registry's array corresponds to an object's subclass. The identifier allows the appropriate singleton to be selected from the registry so that one of its functions can be invoked to perform work in the subclass that the singleton represents.

In other cases, a registry tracks all of the objects derived from a specific class. These objects are created and destroyed at run time. To track an object, the registry assigns it an empty slot in the array. The index of this slot serves as an identifier for the object so that a message, for example, can reference it through its identifier. The registry can detect references to deleted objects by augmenting the identifier with a per-slot **incarnation number**, which it increments when an object is deleted and deregistered.

4.2.4 Polymorphic Factory

The purpose of a POLYMORPHIC FACTORY is to create an object whose subclass is not visible to the user of the factory. It creates such an object through delegation, which preserves layering and partitioning. A polymorphic factory is a form of ABSTRACT FACTORY [GHJV95].

Consider creating an object to represent a card. Say that a card manager is invoked when a craftsperson inserts a card into a slot. The card manager should not `#include` every card-specific `.h` file and create card-specific objects in a `switch` statement. It can avoid the need for such global knowledge by defining `CardRegistry`, a registry for `CardFactory` objects (see Figure 4.1). The purpose of each `CardFactory` is to create the appropriate subclass of `Card` object. Each `Card` subclass defines a `CardFactory` singleton and places it in the `CardRegistry`. This singleton is an example of a CONCRETE FACTORY [GHJV95]. The interfaces look like this:

```
// CardId identifies a card type.
// NilCardId is the nil card identifier.
// MaxCardId is based on the number of card types.
//
typedef int    CardId;
const   CardId NilCardId = -1;
const   CardId MaxCardId = 63;

// Define framework classes for supporting cards.
//
class Card: public Object { }; // details unimportant
```

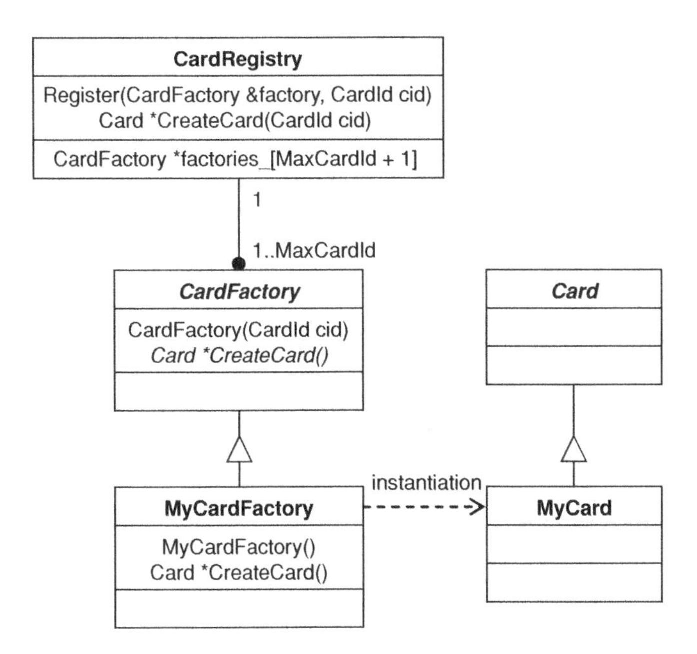

Figure 4.1 Polymorphic factory. `CardRegistry` is a polymorphic factory that creates a `Card` subclass associated with an identifier, `CardId`, by delegating to a `CardFactory` subclass that is associated with `CardId`.

```
class CardFactory: public Object
{
public:
   // Subclasses must provide the Create method. The
   // constructor registers the subclass against CID.
   //
   virtual Card *Create(void) = 0;
protected:
   CardFactory(CardId cid);
};

class CardRegistry: public Object
{
   friend class SingletonObject<CardRegistry>;
public:
   static bool ValidCardId(CardId cid);
   void Register(CardFactory &factory, CardId cid);
   Card *CreateCard(CardId cid);
protected:
   CardRegistry(void);
   ~CardRegistry(void);
private:
   CardFactory *factories_[MaxCardId + 1];
};
```

```
// Use the framework classes to support a new type
// of card, which is identified by SomeCardId.
//
const CardId SomeCardId = 7;

class SomeCard: public Card { }; // details unimportant

class SomeCardFactory: public CardFactory
{
   friend class SingletonObject<SomeCardFactory>;
public:
   virtual Card *Create(void)
                     { return new SomeCard (); };
protected:
   SomeCardFactory(void);
   ~SomeCardFactory(void) { };
};
```

The function implementations for the above classes are

```
CardFactory::CardFactory(CardId cid)
{
   // Add this factory to the global registry of card
   // factories.
   //
   CardRegistry *reg =
      SingletonObject<CardRegistry>::Instance();
   reg->Register(*this, cid);
}

CardRegistry::CardRegistry(void)
{
   int i;
   for(i = 0; i <= MaxCardId; i++) factories_[i] = NULL;
}

CardRegistry::~CardRegistry(void)
{
   int i;
   for(i = 0; i <= MaxCardId; i++)
   {
      if(factories_[i] != NULL)
      {
         delete factories_[i];
         factories_[i] = NULL;
      }
   }
}
```

```
bool CardRegistry::ValidCardId(CardId cid)
{
   return ((cid > NilCardId) && (cid <= MaxCardId));
}

Card *CardRegistry::CreateCard(CardId cid)
{
   // Delegate card creation to the factory
   // associated with CID.
   //
   if(!ValidCardId(cid)) return NULL;
   if(factories_[cid] != NULL)
      return factories_[cid] ->Create();
   return NULL,
}

void CardRegistry::Register(CardFactory &factory,
                            CardId cid)
{
   if(!ValidCardId(cid)) return;
   factories_[cid] = &factory;
}

SomeCardFactory::SomeCardFactory(void):
   CardFactory(SomeCardId) { }
```

During system initialization, `SomeCardFactory` is simply created by

```
SingletonObject<SomeCardFactory>::Instance();
```

When the `SomeCardFactory` constructor calls the superclass `CardFactory` constructor, it is added to the registry. The card manager can now create appropriate subclasses of `Card` objects as follows:

```
CardId cid = SomeCardId; // cid would actually be an
                         // argument
CardRegistry *reg =
   SingletonObject<CardRegistry>::Instance();
Card *c = reg->CreateCard(cid);
```

The `Create` function defined by `CardFactory` is a FACTORY METHOD [GHJV95]. It is a pure virtual function that each concrete subclass (in this example, `SomeCardFactory`) provides to create its particular type of object (in this case, an instance of `SomeCard`).

4.2.5 Cached Result

One technique for improving the performance of an object is to create a member variable for a frequently accessed value and then provide an inline function to return it. This is trivial if the value was a member variable in the first place, but it can also be used when a function calculates the value. In this case, the member variable is a CACHED RESULT.

The value of a cached result, however, should usually be an invariant. If the value changes during the object's lifetime, the overhead and complexity of keeping it up to date may outweigh the gain in performance. For example, say that *A* owns multiple instances of *B* and that *A* needs to provide a function that returns the number of instances of *B* that it currently owns. Although *A* could maintain the count as a cached result and define an inline that returns it, the trade-off is that *A* must update the count whenever an instance of *B* is created or destroyed. If instances of *B* are always created and destroyed through *A*'s interface, updating the count should be simple. However, if instances of *B* are created and destroyed in other ways, maintaining the count could be detrimental. When *B* is created or destroyed, it must now invoke a function on *A* to update the count. This could actually reduce capacity. Furthermore, it is error prone and creates a coupling from *B* to *A* that may be undesirable. If these types of problem start to surface when implementing a cached result, it is better to leave the code alone.

4.3 IMPROVING AVAILABILITY

This section describes object management techniques that prevent outages and therefore improve an extreme system's availability.

4.3.1 Object Pool

By default, the C++ operators `new` and `delete` allocate memory from a heap. In an extreme system, widespread use of a heap is undesirable because it can lead to memory fragmentation. Although the fragmentation is initially moderate, an extreme system remains in service for a long time. Consequently, fragmentation can slowly increase to the point where memory exhaustion becomes a real possibility. To mitigate this risk, the heap may use policies such as best-fit and the merging of adjacent free areas. However, these policies decrease capacity because of their additional processing costs.

A proven technique in extreme systems is to allocate memory during system initialization. Large numbers of memory blocks are created and placed in pools, on free queues. These blocks are allocated from the heap or through some lower-level memory management function. They remain allocated until the system reboots, thereby avoiding an escalation in fragmentation.

At run time, the memory for an object is obtained by dequeuing a block from the OBJECT POOL associated with the object's class. When an object is deleted, its block returns to its queue. A class whose objects reside in pooled blocks overrides operator `new` to obtain a block from the pool and operator `delete` to return it. Consequently, a system can introduce object pools without making any significant changes to application software.

Detailed sketches of classes that implement object pools appear in Section 4.3.3. This section avoids code details in order to focus on the requirements and high-level design.

The size of each pool must be engineered so that each application will have the amount of memory that it requires when the system is running at peak load. The concept of engineered limits for scarce resources is an important one in extreme systems. Memory is a scarce resource, so its usage must be engineered. At peak load, applications should be guaranteed to have enough blocks for their objects. At the same time, no application should allocate far too much memory, because this kind of over-engineering increases memory costs.

Two guidelines simplify the engineering of pool sizes:

1. All subclasses derived from the same framework base class allocate their objects from the same pool.
2. All blocks in a pool are the same size. This wastes some memory because few objects fill an entire block. However, it considerably simplifies the design, something that we will discuss later in this section.

In a switching product, for example, these guidelines allow many pool sizes to be based on a single parameter, namely, the maximum number of simultaneous sessions that a processor will support. This parameter must be configurable because its value, and therefore the number of objects required in the processor, can differ significantly from one customer site to the next.

Although pool sizes are engineered, it must be possible to expand a pool's size without reinitializing a processor. A pool's size may need to grow when adding more users to the system or when it was simply under-engineered. Regardless of the reason, customers of an extreme system will expect the capability to expand a pool's size without causing an outage. This can be readily supported by allocating a pool's blocks in a two-dimensional array. Each array element points to a contiguous memory segment which contains a fixed number of blocks, say 1K of them. If the initial size of the pool is 5K blocks, then only the first 50 pointers are not NULL. Later, additional blocks can be added after the initial 50 segments.

The reverse procedure, that of decreasing the number of blocks in a pool, need not be supported without an outage. Its only purpose would be to rectify the drastic over-engineering of a pool, which is essentially a procedural error. Beyond that, it would only be a cosmetic optimization. It is also difficult to implement, given that blocks from any segments could currently be in use.

Some systems support self-engineering by automatically increasing the size of a pool when all of its blocks are in use. This prevents under-engineering from causing work to fail. However, it can also lead to over-engineering if a faulty application exhausts a pool by creating objects in an infinite loop. Consequently, a self-engineering strategy should expand a pool gradually and impose an upper limit on its size. If the pool reaches this limit, its size must be increased manually.

Each pool must collect statistics to guide engineering and identify problem areas:

- the total number of blocks in the pool;
- the number of blocks that are currently available;
- how many times allocation succeeded;
- how many times allocation failed (because the pool was empty);
- the maximum number of blocks that were simultaneously in use;
- the largest object that was placed in a block.

Pre-allocating memory pools for objects confers a number of benefits:

- It avoids an escalation in memory fragmentation because blocks are allocated in large, contiguous clusters. However, fragmentation occurs in the sense that few objects completely fill the blocks assigned to them.
- It improves capacity because memory for objects is allocated and deallocated with simple dequeue and enqueue operations. This

avoids the typical processing costs associated with a heap, such as best-fit searching and the merging of adjacent free areas.

- Because the size of each pool is engineered, object allocation fails when a pool is exhausted. This prevents a defective application from gobbling all available memory. Pool sizes should be somewhat over-engineered so that, under peak load, they still contain some spare blocks. The exhaustion of a pool then indicates that some application is misbehaving, perhaps by getting into an infinite loop in which it is allocating objects.

  Note that a pool should not exhaust under peak load. If the system receives more work than it can handle, overload controls (discussed in Chapter 10) prevent it from accepting too much work and exhausting an object pool. If an object pool throws a `bad_alloc` exception, applications should not handle it. The exception should instead be caught in the entry function of whichever thread is running, as described in Section 8.1. To recover from the exception, the thread aborts the work currently in progress, which includes recovering all objects associated with that work. The rationale for this approach is that what has almost certainly happened, given that overload controls should prevent pools from exhausting, is that a fault in the application caused it to gobble blocks from the pool that threw the exception. The application's work should therefore be aborted so that these blocks will be immediately recovered.
- Because blocks are allocated in contiguous clusters, a background garbage collector can easily traverse all of the blocks to recover orphans, thereby protecting the system from slow memory leaks. A detailed discussion of this topic appears in Section 8.3.

Note that garbage collection is a background task. Applications delete objects explicitly. In many cases, application frameworks actually delete objects, relieving applications of this burden. The garbage collector is a therefore a final line of defense. If it finds an orphaned object, it generates a log when recovering the block, to highlight a memory leak that needs to be fixed.

Using a foreground garbage collector, as in Java, is undesirable in extreme systems. It involves the use of a heap, whereas memory pools are preferable. Its overhead decreases capacity. Moreover, designing a garbage collector that doesn't freeze the system while conducting a sweep is challenging. It is precisely when a system is running at peak load that it is rapidly allocating objects. And this is the worst time for a garbage collector to take over. Incoming work accumulates while the garbage collector runs, so the system cannot guarantee its response time. The time spent in the garbage collector would be far better spent handling hundreds of transactions, perhaps averting an overload situation.

The design for object pools, as presented above, has one major drawback. Because all blocks are the same size, but are used by all subclasses of the same framework class, they must be large enough to accommodate the largest subclass. If the smallest and largest subclasses have vastly different memory requirements, too much memory could be wasted.

One way to overcome this drawback is to use a separate pool for each concrete class. However, in a system of even moderate complexity, this cure will prove worse than the disease. There are now many more pools, each of which must still be engineered to meet peak load. Say, for example, that a telephone switch defines the abstract class `HalfCall`, where a basic call consists of two `HalfCall` subclasses, one based on protocol of the incoming interface, and the other based on the protocol of the outgoing interface. If the `HalfCall` class hierarchy contains even a modest amount of polymorphism and inheritance, it will contain many concrete classes. Engineering the number of blocks for the `HalfCall` pool is easy: it is simply twice the maximum number of simultaneous calls. Separately engineering the number of blocks for each concrete class is far more difficult. At some peak usage times, the number of instances of class *A* may be significantly more than that for class *B*, whereas the reverse could occur at other times. At some customer sites, *A* may be used more frequently than *B*, whereas the reverse could be true at other sites. The pool for each concrete class will then need to be engineered on a per-site basis, which is both tedious and error prone. Furthermore, the total memory requirements will probably be greater because each concrete class must be engineered for situations in which it happens to be unexpectedly popular during some occurrence of peak usage.

If using a separate object pool for each concrete class is too unwieldy, another solution is required if making the blocks large enough to accommodate the largest subclass wastes too much memory. A reasonable solution is to impose a limit on the size of the blocks, one that accommodates most subclasses. Objects that exceed this limit must then move some of their data into an **auxiliary data block**. These blocks are also allocated in pools, and they come in various sizes, such as small, medium, and large. The class that supports all pooled objects, `PooledObject`, then defines a queue for holding these auxiliary blocks. A large object places some of its data in one or more auxiliary blocks, which it allocates during its instantiation. Moving some data out of the object is usually simple because there is often a large structure, such as an array, that is easily moved.

4.3.2 Object Nullification

Compilers typically offer the choice of building a debug load or a production load. Among other things, a debug load includes or enables optional software that helps to detect common errors, such as accessing a deleted object. A production load omits this software in order to improve performance. During testing, it is highly advisable to use a debug load. Failing to do so runs the almost certain risk of surprises when the software goes to the customer.

When an object undergoes deletion, its `vptr` is continually updated as the chain of destructor calls climbs the class hierarchy. In other words, the object is continually updated to reflect its current subclass. However, the heap manager only resets (nullifies) the object's data in a debug load, when its memory returns to the heap. Nullification is important because it uncovers uses of deleted objects that would otherwise go undetected if the object's data were left intact.

If you are using object pools, you allocate memory for your pools during system initialization. However, this memory never returns to the heap: when a pooled object is deleted, its block simply returns to its pool's free queue.

Consequently, to support debug loads, an OBJECT POOL must provide OBJECT NULLIFICATION. In debug mode, a pool overwrites its blocks with garbage when returning them to its free queue, just as the heap manager does in a debug load. The garbage is some fixed pattern, such as `0xdfdfdfdf`, which causes an exception if mistakenly used as a pointer. The implementation is simple: `ObjectPool::EnqBlock` overwrites the block with the pattern if this behavior is specified for its pool. Nullification is enabled in lab loads but disabled in field loads. In field loads, there should nonetheless be a way to enable it so that it can help to find any references to stale objects in live systems.

After an object's `vptr` is nullified, it is still possible to call the object's nonvirtual functions, which can in turn access its member data. Thus, nullifying objects won't catch all references to deleted objects. However, a fixed pattern such as `0xdfdfdfdf` will detect uses of stale pointer variables and is likely to cause variables such as `enum`s and array indices to go out of range.

4.3.3 Implementing Object Pools

This section outlines the implementation of an OBJECT POOL that can increase its size at run time. It supports OBJECT NULLIFICATION but omits statistics, auxiliary data blocks, and background garbage collection.

First we define the class `PooledObject`. Objects that are managed by a pool must derive from this class:

```
class PooledObject: public Object
{
   friend class ObjectPool;
public:
   virtual ~PooledObject(void) { };
   virtual ObjectPool *Pool(void) = 0; // subclass must
                                       // identify pool
   virtual void PostInitialize() { };  // discussed
                                       // later
protected:
   PooledObject(void) { link_ = NULL; }; // not on
                                         // free queue
private:
   PooledObject *link_; // link when on free queue
};
```

And now for the class `ObjectPool`:

```
const int MaxSegments = 256; // supports up to 256K
                             // blocks in pool

enum Nullification
{
   NullifyNone, // do not nullify deleted object
   NullifyVptr, // nullify vptr only
   NullifyAll   // nullify entire contents
};

class ObjectPool: public Object
{
public:
   virtual ~ObjectPool(void);
   bool SetSegments(int count);
   void Create(void);
   PooledObject *DeqBlock(size_t size);
   void EnqBlock(PooledObject &obj);
   void SetNullification(Nullification n)
         { nullify_ = n; };
protected:
   ObjectPool(size_t size);
   void Nullify(PooledObject &obj);
private:
```

```
   size_t blockSize_;         // size of blocks in bytes
   int    currSegments_;      // current number of
                              // segments in pool
   int    nextSegments_;      // segments to be reached
                              // by Create
   int    *blocks_[MaxSegments];  // the actual blocks
   PooledObject  *freeq_; // queue of available blocks
   Nullification nullify_;    // how to nullify deleted
                              // objects
};
```

The implementation is

```
const int BlocksPerSegment = 1024;
const int BytesPerLongLog2 = 2; // assuming that
                                // compiler aligns
                                // objects on 4-byte
                                // boundaries
const int NullificationPattern = 0xdfdfdfdf;

ObjectPool::ObjectPool(size_t size)
{
   int i;
   blockSize_ = ((size + 3) >> BytesPerLongLog2)
                            << BytesPerLongLog2;
   currSegments_ = 0;
   nextSegments_ = 0;
   for(i = 0; i < MaxSegments; i++) blocks_[i] = NULL;
   freeq_ = NULL;
   nullify_ = NullifyNone;
}

ObjectPool::~ObjectPool(void)
{
   int i;
   for(i = 0; i < MaxSegments; i++)
   {
      if(blocks_[i] != NULL)
      {
         delete[] blocks_[i];
         blocks_[i] = NULL;
      }
   }
}

bool ObjectPool::SetSegments(int count)
{
   if([illegible]
```

```
   if(count >= currSegments_)
   {
      nextSegments_ = count;
      return true;
   }
   return false;
}

void ObjectPool::Create(void)
{
   int i;
   PooledObject *item;
   int size = (blockSize_ * BlocksPerSegment)
                 >> BytesPerLongLog2;
   while(currSegments_ < nextSegments_)
   {
      blocks_[currSegments_] = new int[size];
      currSegments_++;
      for(i = 0; i < size;
          i = i + (blockSize_ >> BytesPerLongLog2))
      {
         item = (PooledObject*)
                  &blocks_[currSegments_ - 1][i];
         item->link_ = NULL;
         EnqBlock(*item);
       }
    }
}

void ObjectPool::Nullify(PooledObject &obj)
{
   if(nullify_ == NullifyNone) return;
   if(nullify_ == NullifyAll)
      memset(&obj, NullificationPattern, blockSize_);
   else
      // nullify vptr only
      memset(&obj, NullificationPattern, 4);
}

PooledObject *ObjectPool::DeqBlock(size_t size)
{
   // To support efficient enqueue and dequeue
   // operations, freeq_ points to the tail item,
   // which then points to the head item. The
   // queue is therefore circular, with the head
   // item second.
   //
   PooledObject *tail;
```

```
   PooledObject *head;
   if(size > blockSize_) throw bad_alloc(); // object
                                            // too large
   tail = (PooledObject*) freeq_;
   if(tail == NULL) throw bad_alloc(); // no blocks
                                       // left in pool
   head = tail->link_;
   if(head == tail)
      freeq_ = NULL; // queue contained one block, so
                     // is now empty
   else
      tail->link_ = head->link_; // queue contained
                                 // >= 2 blocks
   return head;
}

void ObjectPool::EnqBlock(PooledObject &obj)
{
   PooledObject *tail;
   //
   // Requeuing a block already on the free queue
   // creates a mess.
   //
   if(obj.link_ != NULL) return;
   if(nullify_ != NullifyNone) Nullify(obj);
   tail = (PooledObject*) freeq_;
   freeq_ = &obj;  // OBJ is now tail of free queue
   if(tail == NULL)
      obj.link_ = &obj; // queue was empty: OBJ points
                        // to itself
   else
   {
      obj.link_   = tail->link_; // OBJ points to head
                                 // (after tail)
      tail->link_ = &obj; // old tail points to OBJ
   }
}
```

Each subclass of `ObjectPool` is a singleton. Let's define one for messages:

```
// MessageBlockSize is the maximum number of bytes in
// a Message subclass.
//
const size_t MessageBlockSize = 80;

class MessagePool: public ObjectPool
{
   friend class SingletonObject<MessagePool>;
```

```
protected:
   MessagePool(void);
   ~MessagePool(void) { };
};

MessagePool::MessagePool(void):
   ObjectPool (MessageBlockSize) { }
```

During system initialization, the pool is created as follows:

```
MessagePool *pool =
   SingletonObject<MessagePool>::Instance();
pool->SetSegments(48); // pool will contain 48K blocks
pool->Create();         // to allocate (or extend)
                        // the pool
```

Message objects therefore derive from `PooledObject`. The key point is that a `PooledObject` subclass overrides operators `new` and `delete` to dequeue and enqueue blocks in its pool:

```
class Message: public PooledObject
{
public:
   virtual ~Message(void) { };
   void *operator new (size_t objSize);
   void operator delete (void *msg);
   void operator delete (void *msg, size_t size);
   ObjectPool *Pool(void);
   virtual void PostInitialize(void);
protected:
   Message(void) { };  // this is an abstract class
private:
   // member data not relevant to this example
};

ObjectPool *Message::Pool(void)
{
   return SingletonObject<MessagePool>::Instance();
}

void *Message::operator new (size_t size)
{
   ObjectPool *pool =
      SingletonObject<MessagePool>::Instance();
   return pool->DeqBlock(size);
}

void Message::operator delete (void *msg)
```

```
{
   ObjectPool *pool =
      SingletonObject<MessagePool>::Instance();
   pool->EnqBlock(*((PooledObject*) msg));
}

void Message::operator delete (void *msg, size_t size)
{
   ObjectPool *pool =
      SingletonObject<MessagePool>::Instance();
   pool->EnqBlock(*((PooledObject*) msg));
}

void Message::PostInitialize(void)
{
   PooledObject::PostInitialize();
}
```

4.4 SPEEDING UP OBJECT CONSTRUCTION

The cost of constructing objects is the primary overhead associated with OO. This section describes techniques that speed up the construction of objects and therefore improve an extreme system's capacity. However, these techniques only improve capacity modestly, usually by 1% to 3% each. They also introduce some complexity and are therefore best omitted from an initial design.

However, as a system adds new capabilities, its capacity invariably suffers. Customers of extreme systems want new capabilities but simultaneously insist that a new software release not degrade capacity by more than a few percent. Holding the techniques in this section in reserve allows them to be added by a 'tiger team' that is assembled to resolve an unsatisfactory degradation in capacity. In such a situation, tiger-team members will be thankful to have these techniques at their disposal.

4.4.1 Embedded Object

When one object owns another, in the sense that its lifetime spans that of the owned object, it can make that object one of its members. Such containment saves the cost of a separate memory allocation and deallocation for the owned object, whether the heap or an Object Pool handles the allocation and deallocation.

What if the contained object is polymorphic, so that its class is not known at compile time? In this case, the owner can still avoid the separate allocation and deallocation costs by defining a block of memory within its own data area, so that the polymorphic object can be created within this embedded location. If the contained object would otherwise have been obtained from an object pool, making it an EMBEDDED OBJECT improves performance by avoiding the dequeue and enqueue operations used while creating and deleting pooled objects. The block of memory set aside by the owner must be large enough to accommodate the largest polymorphic object, in the same way that blocks in an object pool must be large enough to satisfy the largest subclass that uses them.

An embedded object overrides operator `new` with a placement `new` that takes a reference to the owner object. The implementation returns a pointer to the data area within the owner that is reserved for an instance of the embedded object.

An embedded object also overrides operator `delete` with a simple `return` statement. There is nothing to delete because the memory occupied by the embedded object still exists within the owner. The object's destructor, however, might provide OBJECT NULLIFICATION to detect any usage of the deleted object.

Here is an example of embedding one object inside another:

```
class EmbeddedObject; // forward declaration

const int EmbedAreaSize = 24;

class EmbedderObject: public Object
{
public:
   EmbedderObject(void);
   virtual ~EmbedderObject(void);
   void *EmbedArea(size_t size);
   void SetEmbeddedObject(EmbeddedObject *obj)
         { embedded_ = obj; };
private:
   int field1_;
   EmbeddedObject *embedded_; // object in embedArea_
   char embedArea_[EmbedAreaSize];
   int field2_;
};

class EmbeddedObject: public Object
{
public:
   EmbeddedObject(EmbedderObject &owner);
   virtual ~EmbeddedObject(void);
   void *operator new (size_t objSize);
   void *operator new (size_t objSize,
                        EmbedderObject;
```

```
    void operator delete(void *obj);
    void operator delete(void *obj, size_t size);
    void operator delete(void *obj,
                         EmbedderObject &owner);
    void operator delete(void *obj, size_t size,
                         EmbedderObject &owner);
private:
    EmbedderObject *owner_; // reference to owner
    int field3_;
    int field4_;
};

EmbedderObject::EmbedderObject(void)
{
    field1_ = 1;
    embedded_ = NULL;
    field2_ = 2;
}

EmbedderObject::~EmbedderObject(void)
{
    delete embedded_;
    embedded_ = NULL;
}

void *EmbedderObject::EmbedArea(size_t size)
{
    if(embedded_ != NULL) return NULL;
    if(size > EmbedAreaSize) return NULL;
    return &embedArea_;
}

EmbeddedObject::EmbeddedObject(EmbedderObject &owner)
{
    owner_ = &owner;
    field3_ = 3;
    field4_ = 4;
    owner.SetEmbeddedObject(this);
}

EmbeddedObject::~EmbeddedObject(void)
{
    if(owner_ != NULL) owner_->SetEmbeddedObject(NULL);
}

void *EmbeddedObject::operator new(size_t size)
{
```

```
   return NULL; // if this object must always be
                // embedded
}
void *EmbeddedObject::operator new
        (size_t size, EmbedderObject &owner)
{
   return owner.EmbedArea(size);
}

void EmbeddedObject::operator delete(void *obj) { }

void EmbeddedObject::operator delete
       (void *obj, size_t size) { }

void EmbeddedObject::operator delete
        (void *obj, EmbedderObject &owner) { }

void EmbeddedObject::operator delete
        (void *obj, size_t size,
         EmbedderObject &owner) { }
```

The owner and its embedded object are created as follows:

```
EmbedderObject *e1 = new EmbedderObject();
EmbeddedObject *e2 = new (*e1) EmbeddedObject(*e1);
```

4.4.2 Object Template

The cost of initializing an object can be expensive because of field-by-field write operations and constructor calls down the class hierarchy. If you need to improve performance in such a situation, one approach is to create an Object Template at system initialization time and use it to initialize the object. Here, *template* does not refer to a C++ template class, but rather to a primal instance of the object that is used to block-initialize subsequent instances. An object template is a form of Prototype [GHJV95].

Code that supports template initialization appears in Section 4.4.5, but let's discuss the general design first.

A template is created using `new` when the system initializes, and a pointer to it is saved. At run time, instances of the object are created by invoking a static member function – let's call it `Create` – that mimics the behavior of operator `new`. However, because `new` itself is not used, the compiler does not generate any constructor calls, which saves a lot of time. Instead, `Create` uses the template to block-initialize the object.

A pooled object, for example, overrides operator `new` by returning a block that it dequeues from its OBJECT POOL. Its `Create` function also dequeues a block, but it then constructs the object itself. It does so by accessing the template and writing it into the dequeued block using a block copy operation. If the new object contains many member variables and lies deep within the class hierarchy, the block copy operation saves a lot of time because it eliminates each constructor call in the hierarchy, along with its field-by-field initializations.

There is one problem, however. If some of an object's member variables need to differ from their default values in the object's template, they must be initialized when the object is created. These variables are ones whose initial values can only be determined at run time. To initialize them, a function that resembles a constructor is needed. It differs from a regular constructor, however, in that it only initializes variables that are run-time dependent. Beyond that, it must follow the normal pattern of constructor calls (that is, down the class hierarchy).

`PooledObject::PostInitialize` provides this form of constructor. To mimic the order of constructor calls, a class that requires dynamic initialization calls `PostInitialize` on its superclass *before* it assigns dynamic values to its own member variables.

4.4.3 Quasi-Singleton

Pooled objects are rarely singletons. If they were, we would just allocate them from the heap during initialization and be done with them. However, some pooled objects *often* behave as singletons. If we could treat them as singletons on those occasions, while also handling situations when they do *not* act as singletons, we could reduce the cost of allocating them. This is the topic of this section.

For example, consider a pool of state machines whose transactions run to completion. That is, when a state machine receives a message, it enters a critical region. Within this critical region, the state machine's event handlers process whichever series of events are raised to handle the incoming message. The critical region only ends when the state machine completes its work and is ready to receive its next message. (The rationale for this design is discussed in Section 5.4.)

In the design just described, all event objects, across all state machines, behave as singletons. Because all transactions run to completion, they are processed serially rather than concurrently. Therefore, only one instance of a given event can exist at any time. Events could therefore be implemented as singletons.

However, there are two possible drawbacks to event singletons. First, the reason for implementing an event as an object is so that it can be parameterized, that is, so that it can contain member data that provides additional information about the event. This information would normally be passed as an argument to the event's constructor. If an event is a singleton, it must provide some other function for initializing this information each time that it is used. This is certainly feasible, so it isn't a significant barrier to implementing an event as a singleton.

The second drawback is that there may be times when a state machine wants to save an event. For example, assume that a state machine is processing an event that requires it to perform an asynchronous query to obtain a result. After it sends the query, the state machine leaves its critical region to await a response. When the response arrives, the state machine may then need to resume its processing of the original event. It is desirable, in such a case, to allow the state machine to save the original event before performing the asynchronous query, so that it can reaccess the event and its parameters when it receives the response and resumes execution.

For example, say that an event requests the setup of a session on behalf of a subscriber. To decide whether or not to accept the request, the state machine must check the subscriber's profile to see if the request is allowed. The service node in which the state machine runs caches subscriber profiles so that they will often be available through direct function calls. But when a profile is not in the cache, the service node must send a message to a central database to obtain it. In this situation, the state machine saves the session request event so that it can finish processing it when the profile arrives.

If state machines can save events, events cannot be singletons. Although saving an event is usually the exception rather than the rule, an event is nonetheless unavailable when a state machine is holding onto it. Consequently, events must be allocated at run time.

In a state machine framework, events are the objects most often created and destroyed. Given that events *usually* behave as singletons, is there a way to capitalize on this?

If an OBJECT POOL implements events, the answer is yes. QUASI-SINGLETON describes the behavior of these objects. Code that supports quasi-singletons appears in Section 4.4.5, but let's discuss the general design first.

Each class that wants to behave as a quasi-singleton defines a static pointer member. During system initialization, the pointer is initialized to reference a block obtained from the class's object pool. The class overrides operator `new` by checking the pointer. If it is not NULL, the block is available and is therefore returned for use after setting the pointer to NULL. If the pointer is NULL, however, the previous state machine has not freed the quasi-singleton, so operator `new` allocates a block from the pool.

The class also overrides operator `delete` by checking if the pointer is NULL. If yes, the pointer is set to the block being freed, making it the quasi-singleton. If no, then another instance has become the quasi-singleton, so the block returns to its pool.

Quasi-singletons improve performance usually by avoiding the `DeqBlock` and `EnqBlock` operations that move blocks out of and back into their pools. These operations do a bit more than simply dequeue and enqueue blocks, however. Avoiding them only offers a modest saving each time an object is created or destroyed, but the total savings, over the many objects created or destroyed in this way, can be significant.

4.4.4 Object Morphing

In some situations, an object must change its class after it is created. Consider the following:

- *An object is created before its eventual class is known*. Say that a hardware device is allocated and controlled through a message-based interface. A would-be device user sends a message to request a device. The request creates a state machine that will manage the device. The state machine receives the message and allocates a device from a pool. The pool contains different types of devices, any one of which could satisfy the request. Initially, the state machine represents a virtual device. However, after it allocates a device, it represents a concrete device. At that point, the state machine must change its behavior to support the specific type of device that was allocated. In other words, its class must change.
- *An object changes its role after it is created*. Consider a message object that wraps an incoming byte stream. The message is delivered to a state machine that forms part of a PIPES AND FILTERS pattern [POSA96]. The state machine analyzes the message, determines that it has no work to do, and therefore decides to relay the message to the next state machine in the chain. To save time, the state machine would like to relay the message without copying it. However, incoming and outgoing messages may have different subclasses: an incoming message should be treated as read-only, and only an outgoing message supports the `Send` function. Consequently, the message has to change its class.

In these situations, the standard approach is to free the original object and create a new one. However, this can be cumbersome. The new object must preserve many attributes of the original one, and any pointers to the original object must be fixed up to reference the new one. Deleting the original object and creating the new one also introduces a performance penalty.

OBJECT MORPHING overcomes these drawbacks. It allows an object to change its behavior at run time without having to be copied or having to flip a flag that controls logic within its functions. Morphing is implemented by dynamically changing an object's class, to one that is closely related, without changing the object's location in memory. In the first of our examples, morphing allows the state machine to morph itself from a generic device superclass to a specific device's leaf class. In the second example, morphing allows a state machine to morph an incoming message subclass to its corresponding outgoing message subclass, after which the message can be sent on its way. In both cases, morphing considerably improves performance and simplifies the design.

Code that supports morphing appears in Section 4.4.5, but let's discuss the general design first. To support morphing, the target class defines a static function. It resembles a constructor in that it returns an object of its class. Its argument is an object that will morph to the target class.

The implementation of morphing presents some challenges:

- The object's `vptr` must change to that of the target class. For this purpose, the target class can define a static variable and initialize it to its `vptr` during system initialization by temporarily allocating one of its objects and reading its `vptr`.
- The location of a `vptr` is compiler specific. Some compilers place it at the beginning of an object, but others make it the first data member of the first class, going down the class hierarchy, to define a virtual function. In either case, however, this means that the `vptr` should reside at the top of an object that ultimately derives from `Object`. You should confirm this by using a symbolic debugger that displays the value of an object's `vptr`. Given a pointer to the object, you can use the debugger to look at the object (in hex format) to see where its `vptr` resides. This is important because morphing is easier to implement when a `vptr` resides at the beginning of its object. This is what we will assume in this section, even though it may only be true for objects derived from `Object`. This restriction should present no problems in practice because all non-trivial objects should ultimately derive from `Object`.
- Because a morphed object does not change location, it must be followed by enough slack space to accommodate any additional member data defined by the target class. If the object is allocated from an OBJECT POOL, this is not a problem because both objects come from a common pool, whose blocks are large enough to satisfy all subclasses.

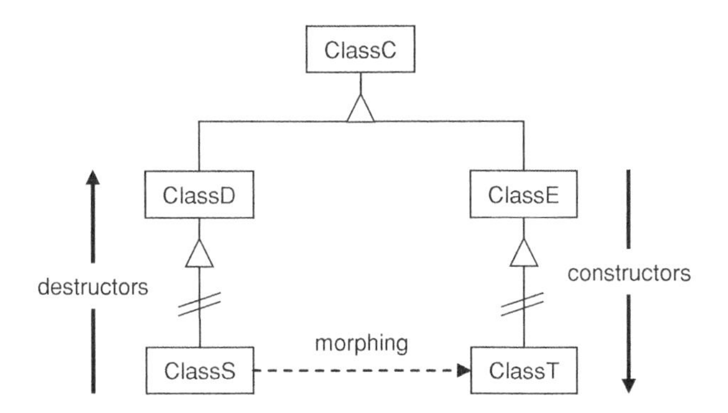

Figure 4.2 Object morphing. To morph an object in ClassS to ClassT, the software must perform the equivalent of destructor calls in the chain ClassS to ClassD, followed by constructor calls in the chain ClassE to ClassT.

- Member variables defined by the target class, but not present in the original object, must be initialized. This can be done by mimicking the behavior of a constructor, with function calls up the class hierarchy until a common superclass is reached. However, the original and target classes usually have the same immediate superclass, in which case this is unnecessary.
- An object has a `vptr` for each of its classes. If an object's class uses multiple inheritance, it therefore has a `vptr` for each base class, and each of them may need to be updated.

Changing an object's `vptr` at run time is unusual – even naughty – but it is faster, and often far easier, than the alternative of performing a deep copy to replace one object with another.

Implementing morphing in a generic way is difficult. Consider morphing an object of class S to an object of class T, where C is the common superclass of S and T (see Figure 4.2). Theoretically, morphing should be implemented by invoking destructors from S up to a direct subclass of C, and then invoking constructors from another direct subclass of C down to T. However, because morphing is used sparingly, in very specific settings, it is reasonable for the target class to contain specific knowledge of any objects that must first be deleted in the source class.

4.4.5 Implementing the Techniques for Pooled Objects

This section illustrates how to support OBJECT TEMPLATE, QUASI-SINGLETON, and OBJECT MORPHING for pooled objects. We begin by defining the class `Cla[illegible]`, which handles many of the details associated with these techniques. Because the techniques occur during object creation, the functions that support them logically belong to a class object, given that classes create objects.

Class is an abstract class. If a subclass of PooledObject wishes to use any of the techniques illustrated in this section, it defines a singleton subclass of Class.

```
class Class: public Object
{
   friend class Object;
public:
   virtual ~Class(void);
   // A subclass overrides SetAttrs to call the
   // protected Set... functions defined below.
   virtual void SetAttrs() {} = 0;
   PooledObject *Create(void);
   PooledObject *GetQuasiSingleton(void);
   void         FreeQuasiSingleton(PooledObject &obj);
protected:
   Class(size_t size);
   void SetPool(ObjectPool &pool) { pool_ = &pool; };
   void SetVptr(Object &obj) {vptr_ = obj.GetVptr(); };
   bool SetTemplate(PooledObject &obj);
   bool SetQuasiSingleton(PooledObject &obj);
private:
   ClassVptr    vptr_;
   PooledObject *template_;
   size_t       size_;
   ObjectPool   *pool_;
   bool         quasi_;
   PooledObject *singleton_;
};

Class::Class(size_t size)
{
   size_ = size;
   vptr_ = NULL;
   template_ = NULL;
   pool_ = NULL;
   quasi_ = false;
   singleton_ = NULL;
}

Class::~Class(void)
{
   delete template_;
   template_ = NULL;
   delete singleton_;
```

```
   singleton_ = NULL;
}

bool Class::SetTemplate(PooledObject &obj)
{
   // Register OBJ as this class's template. The class
   // must already have called SetPool, OBJ's pool
   // must match that pool, and OBJ must not be the
   // quasi-singleton instance!
   //
   ObjectPool *p = obj.Pool();
   if((p == NULL) || (p != pool_)) return false;
   if(&obj == singleton_) return false;
   template_ = &obj;
   return true;
}

bool Class::SetQuasiSingleton(PooledObject &obj)
{
   // Register OBJ as this class's quasi-singleton.
   // The class must already have called SetPool,
   // OBJ's pool must match that pool, and OBJ
   // must not be the template instance!
   //
   ObjectPool *p = obj.Pool();
   if((p == NULL) || (p != pool_)) return false;
   if(&obj == template_) return false;
   singleton_ = &obj;
   quasi_ = true;
   return true;
}

PooledObject *Class::Create(void)
{
   // Create a new object and initialize it using the
   // template.  If the class has a quasi-singleton,
   // use it for the object, else allocate a block
   // from the class's object pool.
   //
   PooledObject *obj;
   if(template_ == NULL) return NULL;
   if(quasi_)
      obj = GetQuasiSingleton();
   else
      obj = pool_->DeqBlock(size_);
   if(obj != NULL)
   {
```

```
      memcpy(obj, template_, size_);
      obj->PostInitialize();
   }
   return obj;
}

PooledObject *Class::GetQuasiSingleton(void)
{
   // If the quasi-singleton is available, return it,
   // else allocate a new block from the class's pool.
   //
   PooledObject *p = singleton_;
   if(p != NULL)
   {
      singleton_ = NULL;
      return p;
   }
   return pool_->DeqBlock(size_);
}

void Class::FreeQuasiSingleton(PooledObject &obj)
{
   // If no quasi-singleton exists, this object
   // becomes the singleton. If a quasi-singleton
   // already exists, return this object to its
   // pool.
   //
   if(singleton_ == NULL)
      singleton_ = &obj;
   else
      pool_->EnqBlock(obj);
}
```

When we defined the `Object` class in Section 4.1, there were some functions that we did not discuss. Two of them support morphing:

```
struct ObjectVptr
{
   void *vptr; // resides at the top of each Object
};

ClassVptr Object::GetVptr(void)
{
   // Return this object's vptr.
   //
   ObjectVptr *obj = (ObjectVptr*) this;
   return obj->vptr;
}
```

```
void Object::MorphTo(Class &target)
{
   // Change this object's vptr to that of the target
   // class.
   //
   ObjectVptr *obj = (ObjectVptr*) this;
   obj->vptr = target.vptr_;
}
```

At the end of Section 4.3.3, we defined the classes `Message` and `MessagePool` as sample subclasses of `PooledObject` and `ObjectPool`. `Message` was an abstract class, so let's define `TextMessage` as one of its concrete subclasses. `TextMessage` will use OBJECT TEMPLATE and QUASI-SINGLETON, so it defines a singleton subclass of `Class`:

```
class TextMessage: public Message
{
public:
   TextMessage(void) {data_ = ""; };
   virtual ~TextMessage(void) { },
   void *operator new(size_t objSize),
   void operator delete(void *msg),
   void operator delete(void *msg, size_t size),
   static TextMessage *Create(void),
   virtual void PostInitialize(void),
private:
   string data_;
};

class TextMessageClass: public Class
{
   friend class SingletonObject<TextMessageClass>;
public:
   void SetAttrs(void);
protected:
   TextMessageClass(void);
   ~TextMessageClass(void) { };
};

// TextMessage uses a quasi-singleton, so it must
// override operators new and delete to support
// the singleton. Note that the singleton is
// private to TextMessage; other subclasses of
// Message inherit operators new and delete from
// Message, which does not define a quasi-singleton.
//
void *TextMessage::operator new(size_t size)
```

```
{
   TextMessageClass *c =
      SingletonObject<TextMessageClass>::Instance();
   return c->GetQuasiSingleton();
}

void TextMessage::operator delete (void *msg)
{
   TextMessageClass *c =
      SingletonObject<TextMessageClass>::Instance();
   c->FreeQuasiSingleton(*((PooledObject*) msg));
}

void TextMessage::operator delete (void *msg,
                                   size_t size)
{
   TextMessageClass *c =
      SingletonObject<TextMessageClass>::Instance();
   c->FreeQuasiSingleton(*((PooledObject*) msg));
}

// The function Create supports template
// initialization. After the superclass Create
// block-initializes the new object, it calls
// PostInitialize to set any fields that depend
// on run-time values or that the template otherwise
// cannot initialize.
//
TextMessage *TextMessage::Create(void)
{
   TextMessageClass *c =
      SingletonObject<TextMessageClass>::Instance();
   PooledObject *obj = c->Create();
   return (TextMessage*) obj;
}

void TextMessage::PostInitialize(void)
{
   // Must invoke superclass first. Note that an
   // object template cannot create a string,
   // because a string is allocated from the heap!
   //
   Message::PostInitialize();
   data_ = "Hello";
}

TextMessageClass::TextMessageClass(void):
   Class(sizeof(TextMessage))
```

```
{
}

// SetAttrs invokes Class::Set functions. Ideally our
// constructor would call these, but it can't because
// some of them require us (as a subclass) to be fully
// constructed!
//
void TextMessageClass::SetAttrs(void)
{
   MessagePool *pool =
      SingletonObject<MessagePool>::Instance();
   SetPool(*pool);
   TextMessage *msg;
   msg = new TextMessage();
   SetTemplate(*msg);
   msg = new TextMessage();
   SetQuasiSingleton(*msg);
}
```

During system initialization, we create `TextMessageClass`:

```
TextMessageClass *tmc;
tmc = SingletonObject<TextMessageClass>::Instance();
tmc->SetAttrs();
```

At run-time, the following creates a `TextMessage` using template initialization. If the quasi-singleton is available, it will house the new object:

```
TextMessage *msg = TextMessage::Create();
```

Now let's look at an example of OBJECT MORPHING. The class `Moth` is abstract. It has two concrete subclasses, `MothLarva` and `MothAdult`, whose objects come from `MothPool`. We plan to morph a `MothLarva` instance to a `MothAdult`, so `MothAdult` defines `MothAdultClass` to support this:

```
class MothPool: public ObjectPool
{
   friend class SingletonObject<MothPool>;
protected:
   MothPool(void);
   ~MothPool(void) { };
};

class Moth: public PooledObject
{
```

```
public:
   virtual ~Moth(void) { };
   ObjectPool *Pool(void);
   void *operator new (size_t objSize);
   void operator delete (void *moth);
   void operator delete (void *moth, size_t size);
protected:
   Moth(void) { };  // this is an abstract class
};

// The details of MothCocoon and MothWing are not
// important.
//
class MothCocoon: public Object { };
class MothWing:   public Object { };

class MothLarva: public Moth
{
   friend class MothAdult;
public:
   MothLarva(void);
   ~MothLarva(void);
private:
   MothCocoon *cocoon_;
};

class MothAdult: public Moth
{
public:
   MothAdult(void);
   ~MothAdult(void);
   static MothAdult *MorphFrom(MothLarva &larva);
private:
   MothWing *wings_[2];
};

class MothAdultClass: public Class
{
   friend class SingletonObject<MothAdultClass>;
public:
   void Initialize(void);
protected:
   MothAdultClass(void);
   ~MothAdultClass(void) { };
};

const size t MothBlockSize = 80;
```

```
MothPool::MothPool(void):
   ObjectPool(MothBlockSize) { }

// The usual overrides of Pool and operators new and
// delete for a class that uses an object pool.
//
ObjectPool *Moth::Pool(void)
{
   return SingletonObject<MothPool>::Instance();
}

void *Moth::operator new (size_t size)
{
   ObjectPool *pool =
      SingletonObject<MothPool>::Instance();
   return pool->DeqBlock(size);
}

void Moth::operator delete (void *moth)
{
   ObjectPool *pool =
      SingletonObject<MothPool>::Instance();
   pool->EnqBlock(*((PooledObject*) moth));
}

void Moth::operator delete (void *moth, size_t size)
{
   ObjectPool *pool =
      SingletonObject<MothPool>::Instance();
   pool->EnqBlock(*((PooledObject*) moth));
}

MothLarva::MothLarva(void)
   { cocoon_ = new MothCocoon(); }

MothLarva::~MothLarva(void)
   { delete cocoon_; cocoon_ = NULL; }

MothAdult::MothAdult(void)
{
   wings_[0] = new MothWing();
   wings_[1] = new MothWing();
}

MothAdult::~MothAdult(void)
{
   delete wings_[0];
```

```
    wings_[0] = NULL;
    delete wings_[1];
    wings_[1] = NULL;
}

MothAdult *MothAdult::MorphFrom(MothLarva &larva)
{
    // To morph LARVA to ADULT, replace the cocoon with
    // wings.
    //
    MothAdult *adult = (MothAdult*) &larva;
    MothAdultClass *c =
       SingletonObject<MothAdultClass>::Instance();
    delete larva.cocoon_;
    larva.cocoon_ = NULL;
    larva.MorphTo(*c);
    adult->wings_[0] = new MothWing();
    adult->wings_[1] = new MothWing();
    return adult;
}

MothAdultClass::MothAdultClass(void) :
    Class(sizeof(MothAdult))
{
}

void MothAdultClass::SetAttrs(void)
{
    MothPool *pool =
       SingletonObject<MothPool>::Instance();
    SetPool(*pool);
    MothAdult *adult = new MothAdult();
    SetVptr(*adult);
    delete adult;  // was only created to obtain value
                   // of vptr
}
```

The following code runs during system initialization:

```
pool = SingletonObject<MothPool>::Instance();
pool->SetSegments(1);
pool->Create();
MothAdultClass *mac =
   SingletonObject<MothAdultClass>::Instance();
mac->SetAttrs();
```

The following code performs the morphing at run time:

```
MothLarva *larva = new MothLarva();
// sometime later...
MothAdult *adult = MothAdult::MorphFrom(*larva);
```

4.5 SUMMARY

- Include capacity as a criterion for evaluating designs.
- The cost of constructing objects is the primary overhead of OO.
- Consider combining small objects. Avoid declaring objects on the stack. Pass objects by reference, not by value.
- Use SINGLETONS and FLYWEIGHTS whenever possible. Create them when the system initializes.
- Use POLYMORPHIC FACTORY to delegate the creation of objects.
- Use EMBEDDED OBJECT, OBJECT TEMPLATE, QUASI-SINGLETON, and OBJECT MORPHING to reduce the cost of constructing objects.
- Define an `Object` class at the top of the class hierarchy to ensure that all nontrivial objects support a core set of useful functions.
- Use OBJECT POOLS to avoid fragmentation, to improve capacity, and to support a background garbage collector that recovers orphaned objects.
- Use OBJECT NULLIFICATION to detect references to deleted objects.

5

Scheduling Threads

This chapter discusses how to schedule threads in an extreme system. It also discusses how to use processes.

Many of the techniques used in commercially available operating systems were originally developed in timesharing systems. An extreme system resembles a timesharing system insofar as both perform different functions simultaneously. However, there are also important differences between an extreme system and a timesharing system. First, the software in an extreme system is either developed internally or carefully selected. It is therefore far more trustworthy than software that users run in a timesharing system. Second, the software in an extreme system remains fixed. Its overall behavior can therefore be studied and characterized before the system is deployed. However, the applications that will run in a timesharing system are not known in advance, so their behavior is hard to predict.

As a key component of an extreme system, the operating system should help the system meet its availability, reliability, capacity, scalability, and productivity requirements. However, it should not be surprising that techniques appropriate for timesharing systems may be less than optimal, or even inappropriate, for extreme systems. And unfortunately, this is the case with what have become the standard scheduling policies in most operating systems.

5.1 STANDARD SCHEDULING POLICIES

Most operating systems use **preemptive scheduling**, which means that the scheduler can perform a **context switch** (scheduling out one

Robust Communications Software G. Utas
 ISBN: 0-470-85434-8 (HB)

thread and running another) at virtually any time. A context switch can occur for various reasons:

1. Most operating systems use **priority scheduling**, which means that the scheduler preempts a thread as soon as a higher priority thread is ready to run. A lower priority thread does not run until all higher priority threads complete their work. The purpose of priority scheduling is to meet scheduling deadlines for hard real-time applications.
2. Many operating systems also use **round-robin scheduling**, which means that the scheduler preempts a thread when it has run for a predefined length of time (called a **timeslice**) and another thread of equal priority is ready to run. The purpose of round-robin scheduling is to prevent one application from unduly delaying other applications of equal priority.
3. A context switch can occur when a thread performs a **blocking operation**. A blocking operation is one that may need to wait for something to happen before it can complete its work. While the operation is pending, the thread is scheduled out and is said to be **blocked**. Blocking operations include
   - waiting for a timer to expire (sleeping);
   - waiting for a message to arrive on a socket;
   - waiting for a disk read or write operation to complete;
   - waiting to acquire a semaphore that another thread owns.

5.2 CRITICAL REGIONS

The applications in an extreme system contend for shared resources, including hardware devices, shared memory and variables, and globally accessible objects. If a thread is scheduled out in the middle of using a shared resource, problems can occur if another thread also uses the resource before the first thread resumes execution. For example, if both threads are writing to the console, the result will be interleaved gibberish. If one thread is deleting an object that another one is using, the result will be an exception or some mysteriously incorrect behavior. Consequently, while a thread is using a shared resource, it is said to be in a **critical region**. During this time, it must prevent other threads from using the same resource. It does so by acquiring a semaphore associated with the resource. If another thread wants to access the resource, it must also acquire the semaphore. However, it will be unable to do so, and will therefore block until the first thread finishes with the resource and releases the semaphore.

5.3 PROBLEMS WITH STANDARD SCHEDULING POLICIES

This section discusses some problems that arise from the standard policies of preemptive, round-robin, and priority scheduling. In subsequent sections of this chapter, we will describe techniques that address these problems.

5.3.1 `p()` and `v()` Considered Harmful

Under preemptive scheduling, a semaphore must guard each critical region. Only when a thread performs a blocking operation does it know the time at which it could be scheduled out. Under round-robin and priority scheduling, a thread can be scheduled out at any time, seemingly at random. It must therefore identify and guard each of its critical regions with a semaphore.

Unfortunately, semaphores are frequently forgotten or added in the wrong place. Their use is highly error prone. Here, Dennis DeBruler eloquently describes the joy of preemptive scheduling and semaphores [DEBR99]:

> As a system integrator of code being developed by several programmers, it seems that my job consists of finding critical region bugs. Most critical region bugs can be found only by subjecting the system to load testing. Because of this need to do load testing, we have the additional challenge of finding and fixing the most difficult bugs late in the development cycle. Also, testing becomes a matter of peeling the onion as you find one unguarded critical region after another. Each one smaller and harder to find. Each one closer to the deadline – or further behind the deadline. Thus we have the joy that the worse the bug, the more layers of management are asking, "Have you found the bug yet?" The really small critical region bugs make it out to the field and become the reason for the "once every month or so in one of the hundred customer sites" bug. These are very hard to even recreate in the lab, let alone find.

If you screen resumes and interview candidates for development roles in extreme systems, you have probably noticed how often experience with multithreaded, intricate thread-safe systems is touted, almost as a badge of honor. Yet if you discuss these experiences with candidates, you invariably learn that they were painful. Avoiding such pain is the motivation for COOPERATIVE SCHEDULING, which is discussed in Section 5.4.

5.3.2 Thread Starvation

A major problem with priority scheduling is **thread starvation**, which occurs when higher priority threads have so much work to do that they prevent lower priority threads from running. An extreme system needs a solution for this problem because each of its threads performs important work. If a thread didn't perform important work, it wouldn't be there! It is therefore dubious to argue that one thread should have *absolute* priority over another. However, it is reasonable for one thread to have *relative* priority over another, so that it receives more of the CPU time when the system is busy. This is the motivation for PROPORTIONAL SCHEDULING, which is discussed in Section 5.7.

5.3.3 Priority Inversion

Under priority scheduling, **priority inversion** occurs when a lower priority thread has acquired a resource (usually a semaphore) that a higher priority thread requires. This prevents the higher priority thread from running and effectively lowers its priority to that of the lower priority thread. Some operating systems address this problem with **priority inheritance**, which temporarily raises the priority of the lower priority thread until it yields the resource [DOUG03]. However, not all operating systems implement this behavior, and others implement it incorrectly when more than one resource is involved.

5.4 COOPERATIVE SCHEDULING

Consider a state machine that starts to process an incoming message. When the transaction ends, the state machine sleeps to wait for its next input, so its thread is scheduled out at that time. During the transaction, there are two choices: let it run to completion, or let the scheduler preempt it at any time.

Allowing transactions to be preempted results in the following outcomes:

- A state machine must protect each of its critical regions individually, making it harder to implement.
- It is easy to miss critical regions, so state machines are error prone.
- Missed critical regions lead to mysterious failures. These are hard to debug, so a lot of time is spent looking for them.

- The scheduler performs a lot of context switching. It preempts state machines during transactions, but state machines also schedule themselves out at the end of transactions.
- Each critical region incurs the cost of acquiring and releasing a semaphore.

Running each transaction to completion eliminates these drawbacks. This makes each transaction one big critical region, so state machines do not have to identify critical regions at a granular level. This makes them easier to implement, less prone to error, and less time consuming to debug. During each transaction, only one semaphore is acquired and released. The scheduler avoids unnecessary context switching because a state machine is only scheduled out when it waits for its next message.

COOPERATIVE SCHEDULING allows an application to decide when its thread can be preempted. An application typically chooses to run until it finishes a logical unit of work, such as a transaction. At that point, the application must be able to take a break before it handles its next unit of work. Each work item runs **locked**, that is, within an extended critical region that prevents the scheduler from preempting one locked thread to run another. One name for this strategy is RUN TO COMPLETION [DEBR95], which highlights the *unpreemptable* nature of each unit of work. This book, however, uses the somewhat more common term *cooperative scheduling* [PONT01], which highlights the need for cooperation among locked threads: each one must eventually relinquish its locked status so that another one can run.

All of the applications developed for an embedded system can be designed to cooperate, unlike those in a timesharing system. This situation should be taken advantage of whenever possible. A fundamental design principle for extreme software is to keep things simple so that nasty bugs will be avoided.

5.4.1 Implementing Thread Locking

There are three ways to support thread locking for COOPERATIVE SCHEDULING:

1. Create a global **run-to-completion lock** (a *lock* is another name for a simple semaphore). This is the easiest solution. All threads that run to completion contend for this lock, so the scheduler cannot preempt one locked thread to run another. A locked thread runs unpreemptably with respect to other locked threads. However, a preemptive scheduler can still preempt a locked thread to

run an *unlocked* thread. Consequently, unlocked threads must not contend for the same resources as locked threads.
2. Disable the scheduler. The usual way to do this is to disable clock interrupts, which prevents the scheduler from running. This provides absolute locking, because not even an unlocked thread will be able to preempt a locked thread. It also improves capacity by suppressing futile invocations of the scheduler. However, it may be undesirable for a reason to be discussed in Section 5.4.3.
3. Modify the scheduler to support a global `locked` flag. Set the flag while running a locked thread, and use it to disable preemptive scheduling. This approach also prevents an unlocked thread from preempting a locked thread. It also allows clock interrupts to serve other purposes.

5.4.2 I/O Threads and Invoker Threads

State machines implement many of the applications in an extreme system. They react to asynchronous inputs (messages) that arrive from external interfaces. But if a state machine runs locked, how does it wait for its next input? And how do you solve the problem of demultiplexing inputs from a single interface to different state machines? In many cases, a single message source generates different types of work or similar work for different state machines.

The solution is HALF-SYNC/HALF-ASYNC [SCH96a], which decouples the handling of I/O from applications. At **I/O level** (that is, in software that performs I/O), an incoming message is simply queued against the state machine that will process it, after which the state machine is placed on a work queue. No application work occurs until the state machine actually runs.

When state machines and I/O are separated, an **invoker thread** services the work queue, and an **I/O thread** performs I/O. An invoker thread runs locked because it executes application software, namely the state machines. However, an I/O thread must also run locked. This is necessary because it places state machines on the work queue, whereas an invoker thread removes them. A critical region must therefore protect the work queue to prevent I/O and invoker threads from accessing it at the same time.

Some systems eliminate I/O threads by receiving messages in invoker threads, but this is undesirable in an extreme system that performs session processing. The primary reason for separating I/O from applications is to allow the system to prioritize incoming work instead of handling it in FIFO order. This is critical for supporting the overload control techniques which appear in Chapter 10.

An extreme system usually has many I/O threads instead of just one. Each I/O thread receives messages on a specific IP port. Using a separate I/O thread for each port ensures that, if an I/O thread dies and has to be recreated, messages do not back up on other ports. It also allows the system to allocate more time to important ports, which would not be possible if a single I/O thread merged inputs from all ports by calling `select`.

5.4.3 Run-to-Completion Timeout

If a work queue contains many entries, how long should an invoker thread run? Should it process all the entries in its queue? This would be undesirable because it could prevent other threads from running for a long time. If an I/O thread were sufficiently delayed, its buffer could overflow, causing messages to be lost. There must therefore be a limit on how long a thread may run locked.

After each iteration through its processing loop – which corresponds to a single transaction in the case of an invoker thread – a locked thread checks how long it has run. If it is approaching the limit, it voluntarily schedules itself out. If it has enough time to make another pass through its processing loop, it continues to run, reducing the amount of time spent on context switching.

How is the size of the limit determined? Here, there are two conflicting forces. On the one hand, the run-to-completion lock acts as a global semaphore among all threads that run locked. The limit's upper bound is therefore determined by the most stringent real-time requirement among these threads. On the other hand, the purpose of the lock is to allow applications to perform each logical unit of work unpreemptably. The limit's lower bound is therefore determined by the time needed to perform the longest logical unit of work.

If a processor runs both hard and soft real-time software, the upper bound could be less than the lower bound! Hard real-time applications typically need to run often, but only briefly, whereas soft real-time applications need to run less often, but usually have much longer transaction times. Thus, it could be that the upper bound, as dictated by the hard real-time software, is 5 milliseconds, whereas the lower bound, as dictated by the soft real-time software, is 10 milliseconds. Now what?

In such a case, the simplest solution is to run hard and soft real-time applications on separate processors. Running them on the same processor adds complexity. Two run-to-completion locks are now needed: one with a short limit, for the hard real-time applications; one other with a longer limit, for the soft real-time applications.

Running hard real-time applications at a higher priority allows them to preempt soft real-time applications but prevents the opposite behavior. Note that a semaphore must now guard each resource shared by hard and soft real-time applications, because the separate run-to-completion locks do not ensure mutual exclusion between hard and soft real-time applications. Now we're on the slippery slope to the widespread need for semaphores. It is therefore preferable to run hard and soft real-time applications on separate processors.

Although a locked thread must schedule itself out voluntarily, it is dangerous to rely on this behavior. An invoker thread executes application state machines, so what happens if application code gets into an infinite loop? If there is nothing to stop this, the system hangs. Consequently, the time limit on running locked must be enforced. However, this is challenging. The run-to-completion lock is a semaphore, so what we want is a semaphore timeout. In most operating systems, however, a semaphore timeout refers to a timeout that occurs when trying to *acquire* a semaphore rather than one enforced on its *release*.

We will now look at three ways to implement the RUN-TO-COMPLETION TIMEOUT:

1. Modify the scheduler.
2. Use a POSIX interval timer.
3. Use a WATCHDOG thread.

Modifying the scheduler. If this option is available, it is the preferred solution. A scheduler that supports round-robin scheduling can easily enforce the timeout on the run-to-completion lock. In the simplest case, the length of the timeslice that provides the round-robin behavior can be set to the value desired for the timeout. Thus, if a locked thread's timeslice expires, the scheduler kills it for running too long. This is effectively round-robin scheduling with the death penalty. The locked thread cannot simply be scheduled out. It is holding the lock, so the lock would first have to be released, but doing so would compromise the critical region that the lock protects, so the thread must be killed.

Using the POSIX interval timer (`setitimer`). The timer is set when a thread acquires the lock and is cancelled when the thread releases it. If the thread runs too long, it receives the signal `SIGVTALRM` when the timer expires. The handler registered against this signal then throws an exception to interrupt the thread's activity. The handling of signals and exceptions is discussed in Section 8.1.

POSIX defines three interval timers, and each generates a different signal when it expires. The timer that generates `SIGVTALRM` is based on the amount of time that a thread has actually run. Another timer is based on raw elapsed time and generates the signal SIGALRM. SIGVTALRM is preferable because it handles a mixture of unlocked and locked threads running under a preemptive scheduler. In this situation, a locked thread can be scheduled out to run an unlocked thread. If this occurs, the locked thread must only be debited for the time it actually ran. SIGVTALRM provides this behavior, whereas SIGALRM does not.

Although POSIX defines interval timers, the following problems arise:

- Not all operating systems support interval timers.
- Some operating systems only support SIGALRM, but not SIGVTALRM.
- Some operating systems do not pass SIGVTALRM to a thread, but rather to the process that created it. The process must then relay SIGVTALRM to the locked thread, perhaps by using SIGUSR1 or SIGUSR2 (user-defined signals) or SIGKILL (to simply kill it). Note, however, that some operating systems use SIGUSR1 and/or SIGUSR2 for their own internal purposes.

In Section 5.4.1 we discussed the possibility of disabling clock interrupts to support locked threads. Now we can see the danger of this approach. Disabling clock interrupts might also disable the SIGVTALRM signal, thus subverting enforcement of the run-to-completion timeout. Only if SIGVTALRM is implemented in some other way, perhaps by an auxiliary clock interrupt, would it be safe to disable the scheduler's clock interrupt.

The conclusion is that, if SIGVTALRM is supported, you can probably find a way to use it for enforcing the run-to-completion timeout. However, you will have to read your operating system's documentation carefully, and the eventual solution is unlikely to be portable.

Using a WATCHDOG *thread*. The first time that this thread runs, it just sleeps indefinitely. When a thread locks, the watchdog thread is signaled to wake it up, and it goes back to sleep for the length of the run-to-completion timeout. When a thread unlocks, the watchdog thread is also signaled, and once again it sleeps indefinitely. However, if the watchdog thread wakes up *before* the thread releases the lock, it kills the thread.

This approach is portable and easy to implement. It does not involve modifying the operating system or dealing with non-compliant implementations of interval timers. One of its drawbacks is that it incurs the overhead of a context switch when a thread acquires and releases the run-to-completion lock. Another drawback is that the scheduler could preempt the locked thread to run an unlocked thread. If this occurs, the watchdog thread's timer continues to run, which could cause the locked thread to be killed – the same problem as with `SIGALRM`. Therefore, all threads should run locked if using a watchdog thread.

5.5 DEALING WITH BLOCKING OPERATIONS

When the work performed during each iteration through a locked thread's loop is bounded by the RUN-TO-COMPLETION TIMEOUT, the application software invoked by the thread does not have to deal with critical regions at all. This highly desirable outcome confers the following benefits:

- It greatly reduces the number of critical region bugs.
- It eliminates the cost of obtaining and releasing semaphores.
- It eliminates the cost of context switches caused by waiting on semaphores.
- A single invoker thread can run all the applications, which minimizes the amount of memory consumed by thread stacks.

However, there is a price to pay for these benefits. Specifically, applications must not use blocking operations – any operation that could cause a locked thread to be scheduled out. A locked thread must only be scheduled out *voluntarily*, in its outermost processing loop. There are two reasons for this:

1. When an application uses a blocking operation, its thread must stop running locked. That is, the thread must release the run-to-completion lock so that other threads can run if it blocks. A blocking operation therefore ends the extended critical region that the lock provides. It therefore risks introducing critical region bugs and the need for semaphores, which is what we are trying to avoid.
2. A single invoker thread runs *all* application state machines on behalf of *all* users. If this thread blocks, its work queue quickly backs up, and the system's response time plummets.

A thread should therefore be able to prevent its applications from using blocking operations. This is the topic of Section 5.5.2. However, not all blocking operations can be avoided, so Section 5.5.1 looks at how to support them. Finally, synchronous messaging, usually in the form of **RPCs** (remote procedure calls), is the most frequent cause of blocking in many systems. It is the topic of Section 5.5.3.

5.5.1 Supporting Blocking Operations by Locked Threads

If there is only one instance of a thread, blocking reduces its throughput to the point where it may not be able to keep up with the work assigned to it. If this occurs, multiple instances of the thread are required so that some of them will be available when others are blocked. Such a group of homogeneous threads is called a THREAD POOL.

An I/O thread blocks when it uses a socket function such as `recvfrom` to wait for a message. For the reasons discussed in Section 5.4.2, an extreme system assigns a separate I/O thread to each IP port. Its I/O threads therefore comprise a thread pool, although not a truly homogeneous one.

The same approach, of offloading blocking operations to a thread pool, applies in other situations. Disk I/O is an example. Generally, an extreme system avoids disk I/O, strongly favoring the use of in-memory databases (of subscriber profiles, for example) to improve its response time. This is especially true for data that invoker threads access, given that the system spends most of its time in these threads.

Nonetheless, there may be situations where disk I/O is required. They can be handled by sending a message to a disk I/O thread (or a pool of them, if disk requests are common enough for a single thread's work queue to back up to the point of causing poor response times). The application that sends the request uses an asynchronous message, just as if it were accessing a remote database. Consequently, its state machine must support a new state when it has a disk-read request pending. A disk-write request, however, rarely requires a new state: the thread simply sends the data and assumes that it will be written successfully. If the data is critical, the application might save the data until it receives an acknowledgment, so that it can reattempt the write operation if it fails. However, the disk I/O thread should support reattempts itself, to relieve applications of this burden.

A thread pool must provide a configuration parameter that allows the number of threads to be engineered. To determine how many threads are required, the system must be modelled or observed under peak load. Statistics must be generated, such as a histogram of response times for requests sent to the thread pool. This helps to determine whether or not adding more threads to the pool will actually improve its response time.

If we are implementing Cooperative Scheduling on top of an operating system with a preemptive scheduler, we need to release the run-to-completion lock when performing a blocking operation. For this purpose, we will define the functions `EnterBlockingOperation` and `ExitBlockingOperation`. When a locked thread uses a blocking operation, it must surround the operation with these functions:

```
EnterBlockingOperation();
   // perform blocking operation
ExitBlockingOperation();
```

These functions release and reacquire the run-to-completion lock, respectively.

5.5.2 Preventing Blocking Operations by Locked Threads

There are two ways to prevent applications from using blocking operations when a thread is locked:

1. If the operating system is modified to support a `locked` flag, as described in Section 5.4.1, this flag can be checked in blocking system calls. If the flag is set, the call fails, either by returning a failure code or by signalling the thread with a fatal error.
2. If the operating system is not modified, each blocking call must be preceded by a call to `EnterBlockingOperation`, which is defined by a base `Thread` class. Before this function releases the run-to-completion lock, it invokes a virtual function, `BlockingAllowed`. The default implementation of this function returns `true`. To prevent blocking operations, a thread overrides it so that it returns `false` or throws an exception. In the former case, we need to modify the code shown above to look like this:

   ```
   if(EnterBlockingOperation())
   {
      // perform blocking operation
      ExitBlockingOperation();
   }
   ```

The usual reason for preventing a blocking operation is that it could block the last thread in a Thread Pool. In the trivial case, where the pool contains only one thread, this enforces a programming model that prohibits blocking operations. For a true thread pool, it reduces the risk of work backing up while all threads in the pool are blocked.

Careful engineering of a pool's size should prevent its last thread from blocking, but if this occurs, and if only *some* transactions perform blocking operations, it is often prudent to deny the blocking operation. The work that the thread is trying to perform then fails, but at least the work that does not block will be processed in a timely manner. An alternative is to dynamically increase the pool's size by creating another thread. This has the benefit of making the pool self-engineering. Its drawback, however, is that the pool could grow without limit if some fault causes threads to remain permanently blocked. Any type of pool that implements self-engineering must ultimately enforce a limit on its size to prevent this type of situation from consuming all available memory.

5.5.3 Prohibiting Synchronous Messaging

Synchronous RPCs are common practice in the computing industry. However, they are anathema to extreme systems because they cause a wide range of undesirable consequences:

- RPCs increase scheduling costs: each one causes a context switch.
- RPCs force the widespread use of semaphores to protect data that is manipulated before and after each RPC. This increases the risk of critical region bugs and again increases the amount of context switching, which can occur whenever a semaphore is acquired. There have been cases of would-be extreme systems that didn't even release semaphores *during* RPCs. They suffered priority inversions, deadlocks, and atrocious latencies. **Latency** refers to how long a message waits in a queue before it is processed. It is the primary determinant of response time because, even under a moderate workload, a message waits in a queue for much longer than it actually takes to process it. An RPC timeout usually occurs after many seconds, so this is how long a thread blocks when a timeout occurs. Work backs up during this time, and the situation is most likely to arise when the system is busy, exacerbating the problem. RPC timeouts increase latency significantly.
- RPCs necessitate the use of THREAD POOLS so that a thread will (hopefully) be available to service a request when its peers are blocked during RPCs. The extra threads consume more memory and slow down the scheduler. The performance of many schedulers is O(n), where n is the number of threads.
- RPCs mask state machine behavior because states are not explicitly defined. There is no *need* to define them because the thread's program counter and stack define the application's underlying state. State machines are not part of the system's object model, which makes it hard to trace requirements through to the code.

- RPCs make it difficult to provide session processing frameworks because responses to RPCs enter the system deep within application code rather than top-down, through the framework.
- RPCs are often implemented behind regular function calls. In a large system, this explicit naming of destination functions creates an undesirable degree of coupling between different software components. It is usually preferable to use the Command pattern [GHJV95] to implement communication between components that can reside on separate processors. We will look at an example of this pattern in Section 9.3.

For these reasons, extreme systems usually prohibit synchronous RPCs. RPCs do, however, have some advantages over asynchronous messaging, and these merit discussion.

First, implementing applications with RPCs is easier than implementing them as state machines. The result of an RPC is an ack (acknowledgment), a nack (negative acknowledgment), or a timeout, and the handling of the latter two is usually identical. On the other hand, an asynchronous state machine must handle any input that could arrive while it is waiting for a response to a request. However, this ability pleases users because it allows them to cancel transactions before they time out. As a PC user, aren't you annoyed by applications that put an hourglass on the screen and refuse to accept cancellation attempts? An hourglass means that someone was too lazy to implement an asynchronous state machine, and instead used a synchronous RPC with a long timeout. RPCs may be useful for prototypes, but production code must avoid them.

Second, when a state machine uses asynchronous messaging, the problem of missing event handlers arises when the designer fails to recognize the possibility of some state-event combinations. This problem is far less common with RPCs, where the next event is always an ack, nack, or timeout. However, it is largely avoided if state machines are designed using SDL diagrams [ITU99], which force designers to specify state machine behavior at a detailed level.

Even if a designer anticipates all state-event combinations, the problem may be one of an explosion in the state-event space. Run-to-Completion Cluster addresses this problem [UTAS01]. State machines in the same cluster communicate with priority messages that are placed in a priority queue. Their invoker thread empties this queue before it accepts any messages from outside the cluster. State machines are still designed as if all messages were asynchronous, but the cluster spares them from having to handle messages that could arrive during transient states that occur during their collaboration. In other words, the cluster mimics the behavior of synchronous messaging. However, clustering is only feasible when the priority messages are known, by design, to be *intraprocessor*, because it otherwise equates to holding a semaphore *during* an RPC, in that the cluster blocks until a remote response arrives. Fortunately, the intraprocessor restriction on priority messages presents few problems in actual practice because state machines that frequently collaborate should be collocated for performance reasons.

Another way of dealing with explosions in the state-event space is to provide a deferred message queue. An application can then postpone the processing of a message by moving it to the deferred queue and returning to it later. However, granting an application this privilege creates two risks: unacceptable latencies, and bugs that cause message reordering. RUN-TO-COMPLETION CLUSTER does not face these risks. First, all priority messages are intraprocessor, so they will be processed quickly, preventing latency problems. Second, a sequence of priority messages is always initiated by a message arriving from *outside* the cluster. The sequence represents a single, logical unit of work that simply happens to be distributed among multiple state machines. It cannot cause message reordering because it avoids the need for applications to defer messages.

It is remarkable that, among those who object to using C++ for capacity reasons, there are many who think nothing of developing a system in which a primary facet of the programming model is the use of synchronous RPCs between a plethora of tasks (processes or threads). Each task typically has a name like `blahmgr`, signifying that it 'manages' the 'blah' capability. To an architect of extreme software, the existence of numerous `blahmgrs` should be a red flag, highlighting a system that is probably in dire need of significant reengineering to address poor capacity, long latencies, and an immature programming model in which tasks are the primary objects and where task boundaries often mirror programmer boundaries (that is, everyone writes their own task).

5.6 THE `Thread` CLASS

This section introduces the `Thread` class, whose main purpose is to implement extreme techniques that involve threads. It also provides THREAD-SPECIFIC STORAGE [POSA00] and a WRAPPER FACADE [POSA00] for native threads.

Details of the `Thread` class appear in Section 8.1.5, after we have finished discussing other thread techniques. This section provides code sketches that illustrate how an invoker thread, I/O thread, and timer thread use `Thread` functions to effect COOPERATIVE SCHEDULING.

```
void InvokerThread::Enter(void)
{
   MutexOn();  // run locked all the time
   while(true)
   {
      item = FindWork();  // dequeue an item from the
                          // work queue
      if(item != NULL)
      {
         ProcessWork(item);
         if(MsecsLeft() < threshold)
            Pause(0);  // let others run
      }
      else Pause(-1);  // sleep until work arrives
   }
}

void IoThread::Enter(void)
{
   allocate a socket and bind it to a port;
   MutexOn();
   while(true)
   {
      // Before waiting for a message, release the
      // run-to-completion lock. After a message
      // arrives, wake the invoker thread if it's
      // sleeping, and then let other threads run.
      //
      EnterBlockingOperation();
      recvfrom(socket, *buffer, bytecount);
      ExitBlockingOperation();
      if(bytecount > 0)
      {
         wrap buffer and put it on an invoker thread's
            work queue;
         invoker->Interrupt();
         Pause(0);
      }
   }
}

void TimerThread::Enter(void)
{
   MutexOn();
   start = ticks_now();
   while(true)
   {
```

```
            timer = dequeue a timer from queue_[currq_];
            if(timer != NULL)
               timer->SendTimeoutMessage();
            else
            {
               // currq_ should actually wrap around at
               // modulo the number of queues. timespent
               // needs to be in msecs; we sleep for 1
               // second, adjusted for how long we ran.
               //
               currq_++;
               timespent = ticks_now() - start;
               Pause(1000 - timespent);
               start = ticks_now();
            }
         }
      }
```

These examples included calls to the following `Thread` functions:

1. `MutexOn` locks a thread. There is also `MutexOff`, which unlocks a thread. These functions increment and decrement a counter so that calls to them can be nested. If `MutexOn` is invoked twice in succession, then `MutexOff` must also be invoked twice before a thread becomes unlocked. Nesting allows different code segments to use `MutexOn` and `MutexOff` to protect critical regions at a granular level. Although most threads choose to run locked all the time, some threads may run unlocked. The application code invoked by such threads uses `MutexOn` and `MutexOff` to protect disjoint critical regions.
2. `MsecsLeft` tells a thread how much longer it may run locked. A thread that holds the run-to-completion lock too long is killed. Consequently, a thread may want to call this function before making another pass through a primary processing loop.
3. `Pause` schedules a thread out for a specified length of time:
   - When a thread has more work to do, it schedules itself out for 'zero time'. This places it at the end of the ready queue, so it resumes its execution after any other ready threads have run. The invoker thread, for example, calls `Pause(0)` when its work queue is not empty but it believes that performing more work might cause it to be killed for running locked too long. The I/O thread does the same after it places a message on a work queue, so that other threads get to run when many messages are arriving.
   - When a thread has no more work to do, it schedules itself out indefinitely. Later, when the thread receives work, another thread

must explicitly wake it up. The invoker thread calls `Pause(-1)` when its work queue is empty.

- When a thread performs work periodically, it schedules itself out for a fixed length of time. The timer thread provides a registry for timer objects and sends a timeout message to an application whose timer expires. If the timers have 1-second granularity, the timer thread calls `Pause(t)` after servicing the timers that have expired, where *t* is 1000 milliseconds minus the number of milliseconds for which the timer thread ran.

4. `Interrupt` wakes up a sleeping thread. This capability is required when, for example, an invoker thread calls `Pause(-1)` when its work queue is empty. When an I/O thread adds an entry to the work queue, it must call `Interrupt` to wake up the invoker thread. If an interrupt mechanism is not available, the invoker thread must frequently poll its work queue, so it can only sleep for a short time.

   Just before it invokes `Pause(-1)`, an invoker thread must record the fact that it has scheduled itself out indefinitely. This allows an I/O thread to determine if an invoker thread is in this state, in which case the I/O thread must wake it.

   If your operating system does not provide a function that corresponds to `Interrupt`, you need to use a condition variable [POSIX03]. The invoker thread waits on the condition variable instead of calling `Pause(-1)`. When the I/O thread adds an entry to the work queue, it signals the condition variable to wake the invoker thread.

   Waking a sleeping invoker thread is similar to LEADER/FOLLOWERS [SCH00]. However, LEADER/FOLLOWERS assumes that invoker threads also perform I/O to avoid the cost of separate I/O threads. When an invoker thread receives a message, it must choose another invoker thread to receive the next message if blocking might occur while processing the current message. An extreme system, however, retains I/O threads for the reasons discussed in Section 5.4.2.

5. `EnterBlockingOperation` and `ExitBlockingOperation` surround any call to a blocking operation. The I/O thread, for example, uses these functions before and after reading a socket. If a thread is locked, these functions release the lock before the blocking operation and reacquire it afterwards.

Recall the cheeky title of Section 5.3.1. On the one hand, `p()` and `v()` – the original names for the functions that acquire and release semaphores in the seminal work [DIJK68a] – are indispensable for protecting critical regions. On the other hand, their usage is so error prone that they should be used sparingly and hidden by higher-level functions whenever possible. In this sense, they are comparable to `goto` statements, which are also indispensable but should only be used by the compiler, in the form of jump opcodes that implement higher-level concepts such as `for` and `while` loops [DIJK68b].

5.7 PROPORTIONAL SCHEDULING

An extreme system must ensure that all of its threads get a chance to perform their work. During times of peak load, payload work (usually the handling of user sessions) should get *most* of the CPU time, because this is what generates revenue. However, other work must also receive *some* time. For example, the system must still respond to administrative commands from craftspeople. There have been cases in which an extreme system gave payload work absolute priority over administrative work. Consequently, the system did not respond to administrative commands during peak load. A craftsperson therefore rebooted the system in the belief that it was hung – even though it was processing payload work without any problems whatsoever.

Each thread in an extreme system performs important work. If it didn't perform important work, it wouldn't be there! In a soft real time system, one thread rarely requires *absolute* priority over another. However, payload threads usually require *relative* priority over other threads. In other words, payload threads should be able to use most, but not *all*, of the CPU time.

This is the motivation for PROPORTIONAL SCHEDULING. It assigns a **faction**, rather than a priority, to each thread. Each faction receives a minimum percentage of the CPU time. Hence the term *proportional scheduling*, to highlight that each faction is guaranteed a portion of the CPU time. A faction might contain only one thread, or many threads might be assigned to the same faction. Factions replace priorities: the scheduler has a ready queue for each faction, and a thread's faction is specified when it is created.

The scheduler implements proportional scheduling using a **timewheel**. The timewheel is a circular buffer in which each entry specifies an interval (in ticks, for example) and a list of factions. The scheduler spends the specified amount of time at each entry, choosing the thread to run from the prioritized list of factions. Within each faction, threads are scheduled in FIFO order, using a standard ready queue. If no faction has a thread that is ready to run, the scheduler advances to the next entry and assigns any remaining ticks to that entry (see Figure 5.1).

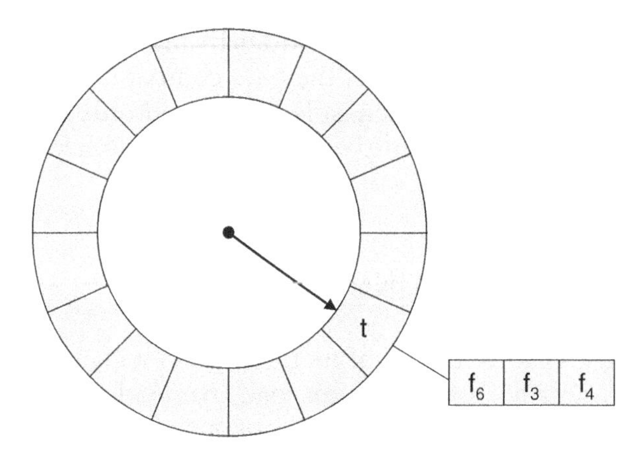

Figure 5.1 Proportional scheduling timewheel. The scheduler will spend *t* ticks at the current entry, running threads in faction 6. If no thread in faction 6 is ready to run, it will run threads in faction 3, followed by threads in faction 4.

Another name for proportional scheduling is PIE SCHEDULING. The timewheel resembles a pie in which some slices are larger than others, with each faction owning a subset of the slices.

5.7.1 Benefits of Proportional Scheduling

Guaranteeing CPU time to each thread faction improves quality and operability by eliminating thread starvation. Using a timewheel minimizes overhead because the scheduler spends very little time on bookkeeping or complex calculations.

In an embedded system, as opposed to a timesharing system, it is far easier to determine each thread's processing requirements and predict its behavior. This characteristic should be used to advantage. Moreover, CPU time is a scarce resource, and an extreme system must engineer the use of scarce resources to ensure optimal performance. Proportional scheduling specifies a scheduler's behavior with high precision, resulting in system behavior that is more predictable than that achievable under priority scheduling.

When a faction has no thread that is ready to run, carrying ticks forward to the next faction allows threads to run at reasonably predictable intervals because each cycle through the timewheel takes the same amount of time. This is useful if some threads require **hard-deadline scheduling**. A remarkable amount of literature exists on this topic [CDKM02]. In the general case, when a system's overall requirements are not well understood, hard-deadline scheduling is a very difficult problem. However, the need for it typically arises in hard real-time systems, whose overall requirements can be known *a priori*. In such settings, the use of proportional scheduling is both a simple and elegant way to meet hard deadlines. In the extreme case, assigning a separate faction to each thread equates to **cyclic scheduling**, which some hard real-time systems use to run each thread at a specific time. Note, however, that carrying ticks forward can cause a faction to run slightly earlier than expected. If this is unacceptable, the leftover ticks could be assigned to some other faction instead.

Under COOPERATIVE SCHEDULING, a thread can only hold the run-to-completion lock for a limited time. Under proportional scheduling, all threads run at regular intervals. Combining strict cooperative scheduling with proportional scheduling eliminates priority inversion because priorities no longer exist. What can occur is **faction inversion**, when an infrequently executed thread owns a semaphore on which a frequently executed thread is blocked. This semaphore must be one other than the run-to-completion lock, and the thread holding it must be running unlocked, so that it is subject to preemption. If faction inversion causes trouble, it can be resolved in the same way as priority inversion, by temporarily changing the faction of the thread that owns the semaphore.

Some schedulers address thread starvation by temporarily raising the priority of threads that have not run for some time. If a scheduler uses this strategy, exactly how the system will behave is hard to predict. Proportional scheduling, however, supports predictable behavior by allowing the system explicitly to engineer its use of CPU time.

When the scheduler provides no defense against thread starvation, a common practice in extreme systems is to run most threads at the same priority. This suggests that the behavior of a proportional scheduler is the desired outcome, even if it can only be approximated in a crude way. If the approximation fails to yield a satisfactory outcome, designers start to play interesting games. One of these is to create additional instances of a thread so that more CPU time is allotted to the function that it performs. Another is to create a high priority thread that monitors other threads and dynamically raises and lowers their priorities in an attempt to achieve the outcome of a proportional scheduler. This approach incurs additional overhead and may again fail to produce the desired result. Once again, proportional scheduling is far more elegant.

5.7.2 Implementing Proportional Scheduling

Unfortunately, few schedulers support proportional scheduling, given that the POSIX standard is blinkered by priority, preemptive, and round-robin scheduling policies [POSIX03]. Consequently, proportional scheduling must usually be added to a system. There are a few ways to do this:

1. If the operating system comes with source code or makes other provisions for modifying its scheduler, direct modification of the scheduler is the best option.
2. If direct modification is impossible, the operating system supplier may be willing to implement the changes on a contractual basis – or even as a strategic investment.
3. The last option is to 'take over the system' with a single native thread that acts as the scheduler. This *One Thread* creates and manages the threads that perform the real work. To perform its scheduling function, it needs the equivalent of a clock interrupt, which the POSIX alarm facility can provide. It must also wrap various operating system calls in a mediation layer that even implements some of them from scratch. If this proves unwieldy, an alternative is to have the One Thread provide a subset of operating system calls that is sufficient to satisfy most threads. Threads that require functions outside of this subset would then be implemented as native threads. The One Thread runs at the highest priority but occasionally gives up the CPU to provide time for these other native threads.

Once a proportional scheduler is in place, the next thing is to determine is which factions are required. Factions are defined based on the types of work that the system performs, examples being payload work, administrative work, and system maintenance (such as reacting to hardware faults). Threads are then assigned to factions based upon the type of work that they perform. Putting each thread in its own faction gives you a lot of control. However, it precludes the sharing of timeslices and forces you to engineer each thread's timeslice individually. It is better to assign threads that perform similar functions to the same faction, particularly if their need for processing time fluctuates. A thread should only have its own faction if it requires hard-deadline scheduling.

After you have assigned threads to factions, the next thing to do is to engineer the timewheel. The timewheel must be implemented using a data structure that can be modified while the system is running. This serves two purposes. First, if you get it wrong, you can fix it without rebooting the system or changing the scheduler code. Second, the system needs the ability to switch autonomously to a different timewheel. During its initialization, for example, the system should suppress payload work until it reaches a stable state, after which payload work receives the majority of CPU time. For example, downloading software to circuit boards should receive a lot of time during initialization but should not interfere with payload work when the system is in service and a new card is inserted. But if an in-service system encounters a flurry of faults, it may switch to a timewheel that increases the time allotted to maintenance activities.

The timewheel must be engineered so that, when the system is running at peak load, it performs the maximum amount of work. The system must exhibit balanced behavior, meaning that important work queues should not back up. Engineering the timewheel therefore involves observing the system's behavior, especially under stressful conditions, and making appropriate adjustments. Although it is easy to define factions, assign threads to them, and adjust the time allotted to each faction, it is harder to determine how much time to allot to each faction.

As an example, a telephone switch might define the following thread factions:

- *Priority*. A ready thread in this faction always runs first. A thread that prevents a hardware WATCHDOG from expiring has a hard deadline requirement and might therefore be the only thread in this faction.
- *Payload*. Examples: processing calls and billing records. CPU time: 85%, because this is what generates revenue. If work in this faction starts to back up, overload controls (see Chapter 10) ensure that the system continues to meet its response time requirements.
- *Provisioning*. Examples: configuring subscriber profiles and hardware. CPU time: 5%. This work is infrequent but should still provide a guaranteed response time.
- *Maintenance*. Examples: handling hardware faults and diagnosing hardware. CPU time: 6% unless the system is in trouble, during which time it might be raised to 20%, with an offsetting reduction in the *payload* faction. This work must run to keep the system healthy.
- *Operations*. Examples: generating logs and other status information. CPU time: 3%. Craftspeople need this information at regular intervals.
- *Audit*. Example: OBJECT POOL audit (background garbage collector). CPU time: 1%. The audit recovers orphaned blocks so that pools do not exhaust.
- *Idle*. A thread in this faction only runs if no other thread is ready. Example: SETI processing. Maybe the system will become famous for finding signs of extraterrestrial intelligence

5.7.3 Approximating Proportional Scheduling

Implementing proportional scheduling is challenging because it means modifying the scheduler or implementing a One Thread to handle scheduling yourself. Consequently, you will probably choose to defer its implementation because you can always introduce it later, with few, if any, changes to application software.

Until you implement proportional scheduling, you can approximate it by running most threads at the same priority to prevent thread starvation. Only threads that you would assign to the *priority* faction need to run at a higher priority. Under COOPERATIVE SCHEDULING, this results in a form of round-robin scheduling where the scheduler cycles through the threads one by one. When a thread schedules itself out, it moves to the end of the ready queue to wait for its next turn to run. All threads therefore get a chance to run, but the percentage of time spent in each thread is roughly the same, assuming that each one schedules itself out when it nears the RUN-TO-COMPLETION TIMEOUT.

The next task, therefore, is to skew the time received by each thread in order to approximate the percentages of CPU time that you would assign to each faction. There are various ways to control a thread's CPU time:

1. Create additional instances of a thread so the scheduler runs its work more often as it cycles through the threads.
2. Alter the amount of time that a thread runs locked. Threads that need more CPU time can schedule themselves out as they near the run-to-completion timeout, whereas threads that require less time can schedule themselves out well before this deadline.
3. If a thread wakes up to perform work at fixed intervals, change its sleep time so that it runs more, or less, often.

To balance the workload during times of peak usage, you must consider the type of work that each thread performs. The *payload* faction, for example, contains an invoker thread, I/O threads, and a timer thread. Let's assume that:

- The run-to-completion timeout is 10 msec (milliseconds).
- One instance of the timer thread is sufficient to send all timeout messages if it wakes up once per second to service all the timers that have expired.
- There are four I/O threads, and the average cost of an I/O transaction (receiving a message on an IP port, wrapping it, and putting it on a work queue) is 0.2 msec.

- There is one invoker thread, and the average cost of its transactions (taking an item from the work queue and running it to completion) is 2 msec.
- Under peak load, I/O threads should receive about 10% as much time as the invoker thread. This rule of thumb minimizes the risk of buffer overflow in I/O threads while allowing the invoker thread to keep up with progress work (see Section 10.1). The percentage is not based on the fact that an I/O transaction costs 10% as much as an invoker transaction, because intraprocessor messages (which are typically progress work) bypass the IP stack (see Section 9.4.2).

Under these assumptions, the invoker thread runs for about 8 msec each time it is scheduled, during which time it handles four transactions. Trying to squeeze in a fifth transaction would be dangerous because the invoker thread might then exceed the run-to-completion timeout and be killed.

In this model, the I/O threads and invoker thread are in balance during times of peak load, assuming that an I/O thread always receives a message when it runs. However, the I/O threads are not efficient. Each one schedules itself out after only 0.2 msec (the cost of one I/O transaction) when it could easily accept as many as 50 messages. To improve efficiency, let's allow an I/O thread to accept 20 messages each time it runs. One cycle through the I/O threads now takes 16 msec (4 threads × 20 messages/thread × 0.2 msec/message), whereas the one invoker thread still only runs for 8 msec. To restore I/O threads to their 10% level, we need 20 instances of the invoker thread (20 threads × 8 msec/thread).

One cycle through the threads now takes 176 msec (16 msec for I/O threads and 160 msec for invoker threads). Therefore, when the timer thread wakes up, it only reaches the head of the ready queue after 176 msec (actually somewhat longer, because of time spent in non-payload threads). This delay is tolerable for timers of 1-second granularity, but you must calculate the cycle time to ensure that this will be the case. If the delay is unacceptable, the timer thread can measure it by comparing the time when it wanted to run with the time when it actually ran. It can then subtract the expected delay from its next sleep time.

5.8 HOW MANY THREADS DO YOU NEED?

An extreme system should restrict itself to a moderate number of threads. The primary reason for having different threads is to partition different types of work. Focusing a thread on one type of work keeps its processing loop simple and prevents the thread's failure from affecting other types of work.

Most threads in an extreme system should be, in UNIX terminology, daemons. They are created during system initialization and are only destroyed if the system is rebooted. Each thread entry function consists of a simple `while(true)` loop that performs the work for which the thread is responsible. Spawning short-lived threads while the system is running is undesirable because it wastes time.

Whenever possible, threads should be designed to be non-blocking to avoid the need for THREAD POOLS. If a primary thread wants to perform a blocking operation, such as I/O, it should use asynchronous messaging to offload the operation to a pool of **worker threads**. These worker threads, however, are also daemons. Creating them during system initialization avoids the cost of creating them when the system is in service, thus improving capacity. Placing them in pools allows the size of each pool to be engineered to prevent too many worker threads from being created and consuming too many system resources.

A recurring design error in would-be extreme systems is implementing ACTIVE OBJECT [SCH96b] using strategies such as THREAD PER USER, THREAD PER SESSION, THREAD PER REQUEST, or even THREAD PER OBJECT. For example, some firms have designed call servers that spawn a thread for each call. To survive, they had to redesign their servers after discovering that running several thousand threads in a processor consumes a lot of memory and causes a major capacity problem. 'THREAD PER...' patterns are rarely appropriate unless threads usually perform blocking operations, as in a server where most transactions must access a disk.

In almost all situations, threads and application objects are orthogonal concepts. Threads do not provide modularity in themselves. They support concurrency, but COOPERATIVE SCHEDULING seeks to minimize this. Consequently, threads in an extreme system are primarily just a way to get CPU time. If application objects avoid blocking operations, a single invoker thread can run all of them.

To summarize, the primary reason for having different threads is to separate different types of work. Multiple instances of a thread are only required when

- Similar work must be performed in more than one faction (or at more than one priority, if PROPORTIONAL SCHEDULING is not used).
- Work involves a blocking operation and is therefore performed by a THREAD POOL to provide an acceptable response time to its clients.
- The operating system does not provide PROPORTIONAL SCHEDULING, so many threads perform the same work to provide it with more CPU time.

- Work would go unperformed for an unacceptable length of time if a thread died and had to be recreated. However, this problem can be largely avoided by catching all exceptions in each thread's entry function, something that will be discussed in Section 8.1.

5.9 SUMMARY

- Adopt Cooperative Scheduling to avoid the error-prone practice of using ubiquitous semaphores to protect critical regions at a granular level.
- The easiest way to implement Cooperative Scheduling is with a global run-to-completion lock that allows a thread to run locked (unpreemptably).
- Enforce a Run-to-Completion Timeout so that an infinite loop in a locked thread will not hang the system.
- Half-Sync/Half-Async separates I/O from applications so that a system can prioritize incoming work. I/O threads put incoming messages on work queues. Invoker threads dequeue work and invoke application objects to process it.
- A Thread Pool provides concurrency when blocking operations are required.
- If applications seldom perform blocking operations, run each transaction to completion by moving all blocking operations from invoker threads to Thread Pools.
- Prohibit synchronous messaging. Write state machines instead.
- Use Proportional Scheduling to avoid thread starvation. Assign threads to factions (types of work) and guarantee each faction a percentage of the CPU time. Use a timewheel to schedule the factions.
- To approximate Proportional Scheduling, use Cooperative Scheduling and run most threads at the same priority. This provides a form of round-robin scheduling. To engineer the percentages of CPU time for would-be factions, create multiple instances of some threads and adjust the length of time that threads run locked.
- An extreme system avoids the cost of creating short-lived threads at run time. Most of its threads are daemons created during system initialization.
- Define a `Thread` class to provide a Wrapper Facade and Thread-Specific Storage for native threads and to implement extreme thread techniques.

6

Distributing Work

This chapter discusses why most extreme systems use distributed architectures. Although distribution offers many benefits, it also introduces complexity, so we will also look at its drawbacks.

The primary question when designing a distributed system is how to partition work among different processors. The final part of the chapter therefore describes different ways to partition work. Its focus is rather high level, taking an architectural perspective rather than delving into low-level details.

Distribution is a primary force in extreme systems. It is therefore a theme that will recur frequently, affecting the detailed design of many techniques that appear in subsequent chapters.

6.1 REASONS FOR DISTRIBUTION

This section discusses the primary reasons why extreme systems are distributed. The purpose of distribution is usually to increase capacity, improve survivability, or to separate applications that require different programming models.

6.1.1 Increasing Capacity

Systems capable of supporting thousands of users with a single processor are rare. Even the fastest processor available might not provide sufficient capacity, perhaps because of poor cache performance. To provide enough total capacity, an extreme system must therefore be distributed. Its design must be scalable so that it can increase its

Robust Communications Software G. Utas
 ISBN: 0-470-85434-0 (HB)

capacity by adding more processors to share the overall workload. The ideal situation is for the system to scale linearly, such that *n* processors yield an *n*-fold increase in capacity. This broadens its target market by making it cost effective at various capacity levels.

Some ways of partitioning work improve capacity more than others. If a system partitions work poorly, adding more processors can actually reduce its capacity. Consider work that is performed with a direct function call. If it moves to another processor, it must be accessed by a message for which a response may also be required. Its total cost therefore increases. Request and response messages must be built, sent, received, and parsed, and there will be two additional context switches. Thus, if the time spent performing the original work is less than the cost of the added messaging and context switching overhead, overall capacity actually decreases. Consequently, a system must distribute sizeable chunks of work to increase its overall capacity.

6.1.2 Improving Survivability

Although the usual purpose of distribution is to improve capacity, it can also improve **survivability**. Survivability refers to a system's *overall* availability. If a processor or software component in a distributed system fails, the others can continue to operate. Although the system may suffer a capacity degradation, it does not suffer an outage. Its survivability – its overall availability – therefore improves. However, this depends on how the system partitions its work. If it assigns specific functions to unique processors, its survivability actually decreases, because the failure of any of these processors or functions renders the function unavailable. Survivability only improves if more than one processor provides each essential function.

6.1.3 Separating Programming Models

In the extreme system reference model of Section 2.2, one reason for having a separate administration node is to improve capacity by offloading the craftspeople's user interface. Another reason is to improve survivability: an administration node does not face extreme requirements, so its applications would be more likely to cause outages if deployed in the core part of the system. Separating the administrative functions offers additional advantages because they can run on a less expensive computing platform, using a non-carrier-grade programming model that is more suited to their needs. These advantages, gained by allowing different types of applications to run in the environments that most suit them, are often compelling enough to motivate the use of distribution by themselves.

6.2 COMPLEXITIES CAUSED BY DISTRIBUTION

Although manufacturers sometimes use distribution as a marketing point, touting a system as 'fully distributed' is missing the point. Distribution is not an end in itself. Its goal is usually to improve capacity or survivability, or both. Once distribution adequately fulfills its goals, it has served its purpose. There is nothing inherently 'better' about a system that is distributed more than is necessary. In fact, *distribution in itself* is undesirable because it complicates software, which makes it more prone to error. Even when distribution improves capacity and survivability, it is not without drawbacks, all of which arise from using messages to perform operations that were previously performed using function calls. We will now look at the complexities that distribution introduces.

6.2.1 Timeouts

The recipient of a message may fail to respond, so the sender must consider the use of timeouts on messages. Timers add further overhead, and the sender must add error-handling code to deal with any timeouts that occur.

6.2.2 Transient States

As discussed in Section 5.5.3, it is highly undesirable to use synchronous interprocessor messages in an extreme system. However, this means that when an application is waiting for a response to a request, it must deal with other messages that could arrive in what is now a new, but transient, state. The application must decide whether to discard, process, or queue these unwelcome messages:

- Discarding the message is a reasonable option when it will be retransmitted, perhaps by user equipment.
- Processing the message is risky because the previous message – the one that caused the request to be sent – has yet to be fully processed. Consequently, incorrect behavior may occur. If the new message aborts the application, however, processing it is usually reasonable

and desirable. The response to the request is then discarded when it arrives.
- In most cases, the message must be queued and processed after the response to the original request arrives.

The general problem is that distribution introduces many more event sequences. This adds complexity, and therefore decreases designer productivity, even when all sequences are taken into account. However, it is easy to overlook some of them, which can lead to decreased reliability or even availability.

6.2.3 Increased Latency

When a system receives a user's request, its response time is primarily determined by how long the request sits in a queue before it is processed. When the request is finally processed, distribution often results in the need for one or more internal (intrasystem) messages to perform some of the work. These internal messages support collaboration and delegation among processors but introduce additional end-to-end delays.

6.2.4 Protocol Backward Compatibility

In a distributed system, upgrading all processors to a new software release simultaneously (that is, in parallel) causes a total outage. A common strategy for overcoming this problem is to upgrade one processor at a time. If the system is distributed in a way that improves its survivability, upgrading its processors serially only causes partial outages or a capacity degradation.

Upgrading processors serially, however, means that there will be times when different processors are running different software releases. This introduces complexity because the processors are still sending each other messages, and the new software release may have changed some of these messages to support new capabilities. These changes, however, must not cause problems for processors that are still running a previous release. In other words, the changes must be upward compatible. Failing that, processors running the new release must be able to revert to the previous message formats when communicating with processors that are running a previous release. However, this means that a processor must know which software release is running in every other processor with which it communicates.

To support compatibility checking, each message header should define a version indicator. This is a frequent requirement in protocols defined by standards, which sometimes go so far as to define a version indicator for each parameter rather than the entire message. The standard then specifies whether a recipient should reject a message or process it to the best of its ability if it encounters a parameter that it does not understand. In Section 9.3, we will see that TLV MESSAGE makes it easy to skip parameters. But although the strategy of rejecting a message may be useful for *intersystem* communication, it is ill suited for *intrasystem* use. A rejected message must revert to a previous version and be retransmitted in any case, which wastes time. It is better to send the message in a format that the recipient will understand, which means that a processor must know which software release is running in every processor with which it communicates.

6.3 TECHNIQUES FOR DISTRIBUTION

This section describes high-level techniques for distributing work, assesses their strengths and weaknesses, and provides examples of their use (see Figure 6.1).

6.3.1 Heterogeneous Distribution

In HETEROGENEOUS DISTRIBUTION, a system distributes functions or applications so that each processor handles a subset of the system's capabilities. Although many systems primarily rely on

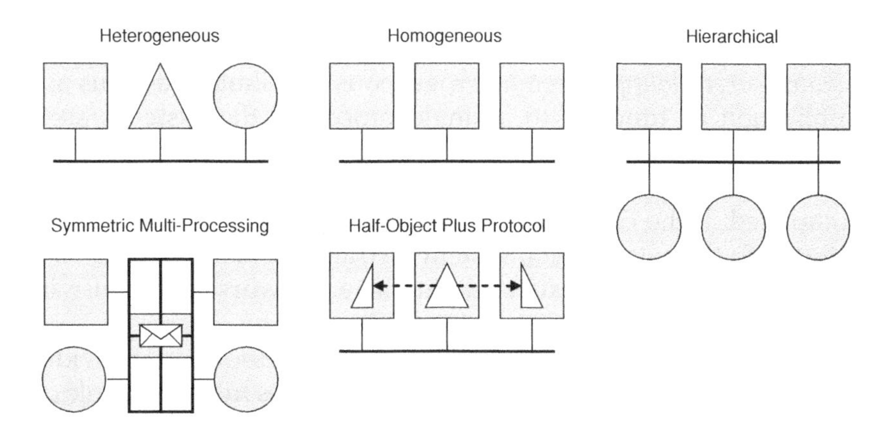

Figure 6.1 Techniques for distribution. In Heterogeneous Distribution, processors perform different functions. In Homogeneous Distribution, they perform the same functions. Hierarchical Distribution mixes these approaches in a Layers pattern. Symmetric Multi-Processing can implement either homogeneous or heterogeneous distribution, but it should use shared memory only to pass messages and to house a database. Half-Object Plus Protocol [illegible] part of an object in other processors to improve their capacity.

heterogeneous distribution, it has a number of drawbacks which limit its use in extreme systems:

- *Limited scalability*. An extreme system usually performs work on behalf of subscribers. Applications running on behalf of a subscriber collaborate to a greater degree than applications running on behalf of different subscribers. A telephone switch, for example, provides many services, whose logic often depends on the attributes and state of an individual subscriber. In most cases, services like call waiting and conference calling can be implemented within a single subscriber's domain, without the need for cooperation from software running on behalf of other subscribers. Although there are exceptions to this rule, software running on behalf of different subscribers usually collaborates only to set up basic point-to-point calls.

  Under heterogeneous distribution, many collaborations that were formerly intraprocessor become interprocessor, given that the applications running on behalf of an individual subscriber are now distributed over multiple processors. Scalability therefore suffers. As previously mentioned, when work moves to another processor, its cost must be greater than that of the newly introduced messaging and context switching overhead. This factor ultimately limits how much work can be distributed heterogeneously. Scalability quickly peaks, and any further attempts at heterogeneous distribution actually cause an overall loss of capacity. Highly scalable systems must minimize their interprocessor messaging, but heterogeneous distribution usually increases the frequency of such messages.
- *Limited survivability*. When heterogeneous distribution assigns an application or function to a single processor, the system's survivability does not improve because the failure of that processor renders the capability unavailable until some form of recovery is completed. If the capability is essential, its loss equates to a total outage. To the extent that a system distributes essential functions to unique locations, its survivability actually worsens because it then contains many single points of failure.

  To improve survivability, more than one processor must provide each essential function. However, this introduces further complexity when deciding how to map functions to processors.
- *Bottlenecks*. If heterogeneous distribution assigns a function to a single processor, that processor can become a bottleneck that limits the system's overall capacity. Identifying bottlenecks and figuring out how to distribute their centralized functions is one of the primary ways to improve scalability.

- *Multiple software loads*. Different processors perform different functions, so the tendency is to build a custom software load for each processor, one that only contains the functions assigned to it. This practice usually ends after everyone realizes what an administrative burden it imposes. It is far easier to build a single software load in which each function can be selectively enabled or disabled. In a large heterogeneous system, however, this simplification can waste a lot of memory in each processor, memory that is occupied by code that never executes.
- *Poor cohesion*. This drawback results from the many collaborations required between applications running on behalf of an individual subscriber, but on different processors. Where is a subscriber's state information located? It quickly becomes distributed across many processors, each of which knows part of the overall picture. In some situations, this creates the need for large messages to export a subscriber's state information from one processor to another, so that an application running in one processor can make decisions based on state information contained in another. In more acute situations, this degenerates into, 'Send me everything you know in case I need to use it'. Even in the early stages of this scenario, the system's capacity degrades. Moreover, all of the complexities inherent in distribution quickly escalate. The software must contend with more and more timeouts and transient states, the system's latency increases, and an ever-increasing number of complex protocols must evolve in an upward compatible manner.

An example of poor cohesion in the telecom industry is the current state of intelligent network (IN) standards. IN allows new services to be implemented outside of telephone switches, in processors known as service control points (SCPs) that are under the direct control of operating companies (telcos). The goal of IN is to allow a new service to be deployed more rapidly, because telcos do not have to wait for each switch supplier to develop it. Instead, the service is developed once and deployed in an SCP. To this end, IN defines ways that an SCP can register with a switch to observe the progress of calls – an example of OBSERVER [GHJV95]. When a call satisfies certain criteria, known as triggers, the switch sends a query to the SCP, asking it for instructions on how to handle the call. The query exports enough of the call's state information to allow services to run in the SCP.

This strategy initially worked well because it defined a focused protocol that allowed certain types of services (involving number translation, call routing, and call screening) to be developed in SCPs. An SCP registered as an observer so that, at various points during call setup, the switch would consult the SCP, which responded by telling the switch to handle the call in various predefined ways: continue as usual, abort the call, route it to a different destination, and so forth.

Given its initial success, it was inevitable that attempts would be made to extend the IN protocol to enable a wider range of services. The initial set of services was limited to those that ran before a call was established, but now the goal was to extend this to services that ran *after* a call was established. This, however, significantly increased the amount of state information that the switch had to export to the SCP. Furthermore, when a switch is waiting for an SCP's response, it is in a transient state. During call setup, the only event that it can receive is a release from the calling user. This simply ends the call, even during a transient state, but once the called user answers, the number of events that can occur during transient states increases significantly. For example, a subscriber might try to set up a conference while the SCP is deciding how to handle a previous event. The need to deal with such complexities made it evident that IN standards had run their course. They readily supported services that ran before a call was answered, but they were not well equipped to support services that ran afterwards.

Although heterogeneous distribution has many drawbacks, there are situations in which it is appropriate. The first of these involves centralizing a capability that is too unwieldy to distribute, such as a large database. A specific example of this is the routing of toll-free calls, one of the many services successfully implemented using IN standards. Unlike most telephone numbers, a toll-free number (an 800 or 888 number in North America) is not associated with a specific interface. When someone places a call to a toll-free number, the actual destination is determined by consulting a database that maps the number to a 'real' number, often based upon the caller's geographical location. This database is so large that replicating it in all telephone switches is infeasible. It therefore resides in a limited number of centralized databases. When a switch receives a request to set up a call to a toll-free number, it sends a query to one of these databases, which responds with the 'real' number to which the call should be routed. This is an example of heterogeneous distribution within a network rather than within a system.

Heterogeneous distribution allows applications to customize their environments by using the platforms and programming models that are most suited to their needs. Early in this chapter, this was mentioned as one reason for having a separate administration node. In a similar vein, heterogeneous distribution serves to separate software with hard real-time requirements from software with soft real-time requirements, the need for which was discussed in Section 5.4.3. Although this enforces a LAYERS pattern [POSA96], its primary purpose is to simplify scheduling. Large IP routers and telephone switches are often designed in this way, being partitioned into the service nodes and access nodes of Section 2.2. High-level logic resides in service nodes, which set up connections between subscribers. However, subscribers interface to the system through access nodes, which provide low-level protocol processing, such as scanning hardware and sending and receiving messages to and from subscriber interfaces. Once the service nodes establish a connection between two subscribers, the access nodes handle the subsequent user-to-user data flow, whether in the form of voice or packets. Service nodes have soft real-time requirements, whereas access nodes have hard real-time requirements.

6.3.2 Homogeneous Distribution

If HETEROGENEOUS DISTRIBUTION can be summarized as, 'Move some *functions* to another processor', then HOMOGENEOUS DISTRIBUTION can be summarized as, 'Move some *subscribers* to another processor'. This approach usually avoids the pitfalls inherent in heterogeneous distribution.

- *Scalability*. Large networks, such as IP or telephone networks, are primarily based on homogeneous distribution. They assign a subscriber to a specific system, which runs most of the subscriber's services. Apart from setting up connections to other subscribers, most collaboration occurs within each system, among the various services running on the subscriber's behalf. Consequently, the network scales linearly. To support more subscribers, you just add another system. Given that the network itself is designed in this way, there is a compelling case for using the same approach to distribute work among the nodes that are *internal* to a system. This is an example of RECURSIVE CONTROL [SEL98].
- *Survivability*. The failure of a processor only affects some subscribers. These subscribers experience a service outage, but the system as a whole only experiences a partial outage.
- *One software load*. Each processor performs the same functions for a subset of the system's subscribers, so each of them runs the same software load.
- *Cohesion*. Collaborations usually occur within a single processor, among the various applications that run on a subscriber's behalf. Interprocessor collaborations are limited to those that occur between applications running on behalf of different subscribers. This

set of interprocessor collaborations is *minimal* and *unavoidable*, assuming that a single processor cannot serve all subscribers.

The design of a resource pool for hardware devices illustrates some trade-offs between homogeneous and heterogeneous distribution. The heterogeneous approach uses one central pool. An application that needs a device sends a request to this pool, and the response assigns one. When the application no longer needs the device, it returns the device to the pool by sending a release message. The homogeneous approach, on the other hand, assigns a subset of the devices to each of the processors whose applications use them, thereby creating a set of local pools instead of one central pool.

The disadvantages of a central pool (the heterogeneous design) are that

- Messages that allocate and release devices decrease capacity.
- Applications must support a new state when a device request is pending.
- If the processor running the central pool fails, it affects all device clients.

The disadvantage of local pools (the homogeneous design) is that the system usually needs more devices. It must provide enough devices to handle times of peak usage in each processor *individually*. With a central pool, the number of devices only has to handle times of *system-wide* peak usage, and this peak is usually less than the sum of the possible peaks within each processor.

In practice, the way that software implements device pools almost always mirrors how the hardware architecture distributes the devices. If the devices reside in a separate shelf or on a separate card, a central pool manages them. But if they are dispersed and collocated with the processors whose applications use them, local pools manage them.

6.3.3 Hierarchical Distribution

Most extreme systems combine heterogeneous and homogeneous distribution. This can be referred to as HIERARCHICAL DISTRIBUTION. Horizontally (with respect to layering) the distribution is homogeneous, but vertically it is heterogeneous. The reference model introduced in Section 2.2 is an example of hierarchical distribution. The administration, control, service, and access nodes provide a heterogeneous distribution of functions, whereas peer service nodes and peer access nodes provide homogeneous distribution.

6.3.4 Symmetric Multi-Processing

In SYMMETRIC MULTI-PROCESSING (SMP), a group of processors shares a common memory. The same software is loaded into each processor and runs in parallel. Bringing more processors to bear on the problem increases capacity.

For the types of extreme systems that are the topic of this book, SMP is a case of the hardware team having fun at the expense of the software team. The principal problem is that SMP forces software to handle critical regions at a very granular level so that one processor will not cause all the others to block. The outcome is the highly error prone type of software that we sought to avoid by using COOPERATIVE SCHEDULING. Classic SMP precludes cooperative scheduling because software running in parallel mimics the characteristics of preemptive scheduling, where each critical region must be protected individually.

Even when a system *does* handle critical regions at a granular level, it may still experience a lot of blocking on semaphores. SMP systems rarely scale linearly as processors are added. Blocking often causes them to reach a capacity asymptote rather quickly.

Finally, as the saying goes, there are only three good numbers: 0, 1, and ∞. In other words, will four processors (or eight, or whatever) always provide enough capacity for the entire system? If not, then why bother with SMP? Another form of distribution will be required anyway, and there is no prize for doing the same thing two different ways.

There may be compelling reasons for implementing an extreme system on an SMP platform, but they have nothing to do with software. Perhaps the SMP platform is cost effective, or perhaps it reduces the physical size of the system when having a small footprint is critical. If you face this situation and need to implement an extreme system on an SMP platform, keep it simple. Given n processors, divide the memory into $2n$ segments and assign two segments to each processor: one large and private, and one small and shared. Use the small shared segments to pass messages between processors, just as if they were physically separated, but run all work in the large private segments. In this way, only the messaging system has to worry about granular critical regions, and it can typically use spinlocks [DOUG99] for this purpose, while waiting for access to another processor's shared messaging segment. The result will be a system that is far more reliable, and probably far more scalable, than one designed using classic SMP. The processors could all run the same software (HOMOGENEOUS DISTRIBUTION), different software (HETEROGENEOUS DISTRIBUTION), or a combination of both (HIERARCHICAL DISTRIBUTION).

6.3.5 Half-Object Plus Protocol

Extreme systems frequently use HALF-OBJECT PLUS PROTOCOL (HOPP) [MESZ95] to alleviate some of the effects of distribution. HOPP encompasses a variety of situations in which the master instance of an object is fully or partially replicated in other processors.

As an example, consider the setup of a call between two telephone subscribers who are both served by the same switch. The switch receives a setup request from the calling user and eventually decides to establish a call to the called user. If different processors serve the two users, the first processor must send the setup request to the processor that serves the called user. How can it identify the correct destination processor?

One approach is to use HETEROGENEOUS DISTRIBUTION. The calling processor sends a request, containing the called user's address, to a name server. This server responds with the address of the processor that serves the called user. This is certainly a reasonable design, one that is analogous to the use of DNS servers in IP networks. However, it does incur an overhead of two messages per call. HOPP and HOMOGENEOUS DISTRIBUTION can eliminate these messages by replicating the name server's database in all of the processors so that they can perform destination lookups locally. HOPP therefore alleviates some of the effects of distribution:

- *Messaging costs*. In our example, a direct function call to the replicated object (the name server database) replaces two messages (the query and response involving the name server). This increases the call-handling capacity of each processor.
- *Timeouts*. Eliminating a query–response message pair relieves applications of the need to deal with message timeouts when the remote object fails to respond.
- *Latency*. Replacing messages with function calls reduces latency. There is no turnaround time introduced by waiting for a response from a remote object.
- *Transient states*. Replacing messages with function calls eliminates transient states. Applications need not deal with messages that can arrive while the work associated with previous message is still pending, awaiting a response from a remote object.

HOPP also has some drawbacks, however:

- *Memory usage*. Fully or partially replicating an object requires more memory, the cost of which may be prohibitive.
- *Messaging costs*. Propagating changes in a primary object to its remote instances introduces more messages.
- *Transient states*. Until HOPP propagates a change in a primary object to a remote object, an application using the remote object can obtain the 'wrong answer'. Although this only occurs during a brief interval, it may nonetheless cause an unacceptable error in the application. The frequency with which HOPP propagates changes determines the length of the interval, but HOPP cannot eliminate it completely.
- *Complexity*. Adding protocols to fully or partially replicate objects increases the system's complexity.

Before using HOPP, weigh its advantages against its drawbacks. Although HOPP usually comes out ahead, this outcome is not a certainty:

- HOPP's memory costs may be too great.
- It may be unacceptable for an application to obtain a "wrong answer" during the interval in which a remote object contains stale data.
- A careful analysis may reveal that HOPP adds more messages (object updates) than it eliminates (object queries). This by itself, however, should not necessarily preclude the use of HOPP, because its offsetting benefits – reduced latency and simplified application software – are compelling.

6.4 SUMMARY

- Extreme systems use distribution to provide scalability, improve survivability, and separate applications that require different programming models.
- Distribution complicates software by introducing timeouts, transient states, and the need for PROTOCOL BACKWARD COMPATIBILITY. It also increases latency and introduces messaging costs.
- HOMOGENEOUS DISTRIBUTION assigns different subscribers to processors. It is often the best strategy for distribution.
- HETEROGENEOUS DISTRIBUTION assigns different functions to processors. Its purpose should either be to centralize a function that is difficult to distribute or to separate applications that require different programming models.

- Hierarchical Distribution uses Homogeneous Distribution within a layer and Heterogeneous Distribution between layers. The latter typically serves to separate soft real-time software from hard real-time software.
- Symmetric Multi-Processing (SMP) can implement any of the above forms of distribution. Its processors should only use shared memory to exchange messages or to share a database that is read-only except to the processor that manages it. Classic SMP is undesirable in an extreme system because it reintroduces the need for ubiquitous semaphores.
- Half-Object Plus Protocol (HOPP) often improves capacity by partially or fully replicating data in other processors so that they can access it locally.

7

Protecting Against Software Faults

An extreme system experiences software faults because of the size of its code base and the number of programmers contributing to it. The topic of this chapter is how to prevent faults from causing further harm.

One example of a software fault is generating and using a bad pointer. Writing through a bad pointer can cause a data corruption, so this is something that we must try to prevent. Many of the techniques in this chapter serve this purpose, but if they fail to prevent a corruption, we then face an error that could lead to a failure. The next chapter (Chapter 8) will therefore discuss how to recover from faults that actually lead to errors.

7.1 DEFENSIVE CODING

Perhaps as much as 90% of the software in an extreme system involves anticipating, detecting, and handling errors. The following types of things are standard practice in extreme software:

- Check that a pointer or array index is valid before using it.
- Check that the arguments supplied to a function are valid.
- Check a function's result to see if it failed, and handle any failure as gracefully as possible.
- Check that a message arriving on an external interface is properly formatted.

Robust Communications Software G. Utas
 ISBN: 0-470-85434-0 (HB)

- Start a timer when sending a message that could lead to a hung resource if the destination does not respond.

It is up to software developers to practice this type of DEFENSIVE CODING, so you need to make it part of your development culture. Include defensive checks whenever there is a risk of causing an exception, trampling data, orphaning resources, or of causing unpredictable behavior. A primary purpose of code inspections must be to ensure that software adequately deals with these risks.

An interesting dynamic in software teams is that defensive checks often reflect programmer boundaries. Everyone trusts their own code, so they only include defensive checks when interacting with other people's code. Of course, it is good practice to include defensive checks even when interacting with your own code.

Because defensive coding carries a real-time penalty, try to avoid redundant checks (in nested functions, for example). You can also disable expensive checks in the field after proving the software in the lab, although this is something to consider carefully. 'This could never happen' are often famous last words in a system that processes millions of transactions every day. Thus, although you might *disable* a defensive check in the field, you must not *eliminate* it with an `#ifdef`. Instead, bypass it using a boolean that you can enable when you need to reinstate the check to guard against unexpected behavior.

7.2 PROTECTING AGAINST TRAMPLERS

A **trampler** is an application that corrupts memory, usually because it uses a faulty pointer or array index. Corruption can also occur as the result of critical region bugs, although COOPERATIVE SCHEDULING minimizes this risk.

A trampler is difficult to find because the corruption that it causes rarely results in an immediate outage. The outage usually occurs later, long after the trampler has left the scene of the crime. Tramplers are therefore the most feared problem of all, and an extreme system must go to some length to guard against them. During testing, tools like Purify help to detect tramplers, but in a large system, tramplers occasionally escape into the field, and the system must be ready for them. To this end, there are a number of techniques, beyond simply checking pointers and array indices in application code, that merit consideration. The following sections discuss these techniques, each of which protects some portion of memory (see Figure 7.1).

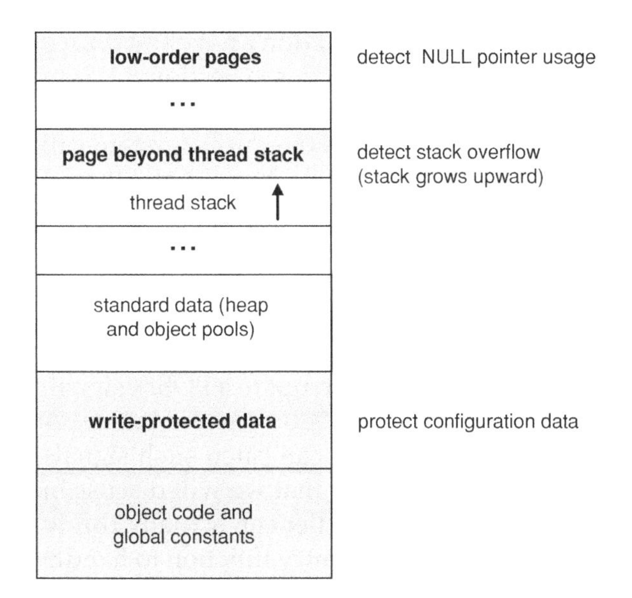

Figure 7.1 Memory protection. Most systems protected their object code and global constants. An extreme system must also protect itself against stack overflows, NULL pointers, and the corruption of its configuration data.

7.2.1 Stack Overflow Protection

If a thread overruns its stack, by nesting function calls too deeply or creating too many local variables, it tramples the memory that lies beyond its stack. The operating system must prevent stack overruns. If it doesn't, you must either change it to do so or discard it in favor of one that does.

The first way to prevent stack overruns is to avoid declaring large local variables, particularly large arrays. This is something to watch for during code inspections. If a large structure is required, allocate it dynamically, either as an object or by using `malloc`. The function must also ensure that it frees the memory before returning, to prevent a memory leak.

Recursion can also cause stack overruns. There are situations where recursion is legitimate, but you must analyze them to ensure that excessive nesting will not occur.

Avoiding large local variables and deep recursion minimizes the risk of stack overruns. However, the system must still guard against them in case they occur. The usual way to do this is to allocate a protected page at the end of the stack. If the stack overflows, the attempt to write to this page causes an exception. If the operating system doesn't do this itself, you should still be able to implement it. If you have the operating system's source code, you can modify it to add the page during thread allocation. If you don't have the source code, you can modify the `Thread` constructor (see Section 8.1.5) to add two pages (for rounding purposes) to the size of each thread's stack. In the `Thread` entry function, you can then use a system call to write-protect the topmost page in the stack. Now, if the thread overruns its original stack size, an exception occurs.

Well, not exactly an exception, at least not in the C++ sense. In most systems, what happens is that the CPU (or memory management unit) generates what is effectively an interrupt when it detects a bad memory access, whether to memory that is either off-limits or non-existent. The operating system maps this interrupt to a POSIX signal, either `SIGSEGV` (in the former case) or `SIGBUS` (in the latter). A signal handler registered against the thread can catch such signals and initiate recovery procedures, something that we will discuss in Section 8.1. In some systems, this signal handler can actually throw a C++ exception, which allows the `Thread` entry function to handle everything that would otherwise be fatal.

However, it gets messier. When a stack overflow occurs, it is vital that there actually be space on the stack to throw an exception! The signal handler also runs on the stack, and we're already at the end of the stack, or at least very close to it. Furthermore, C++ constructs exceptions on the stack. So if the signal handler uses local variables or throws an exception, it will probably cause another `SIGSEGV` itself, for once again overrunning the stack. Depending on the operating system, this will result in a crash or an infinite loop. Now what?

In some operating systems, the signal handler can request a separate stack for its execution, using the call `sigaltstack`. The question is whether an exception thrown from this stack will find its way back to the original stack. The answer depends on the system's choice of compiler and operating system.

If it is impossible to ensure that the signal handler can successfully throw an exception or otherwise recover from a stack overflow, the thread must instead be killed and recreated. Recreating a thread is the responsibility of the parent thread (the thread that originally created it). Most operating systems signal the parent thread (using `SIGCHLD`) when killing one of its children. If the operating system does not provide this capability, the parent must frequently wake up and check the status of its child threads. We will return to this topic in Section 8.1.

7.2.2 Low-Order Page Protection

Unfortunately the value of NULL is 0, which means that using a NULL pointer (perhaps with an offset) causes scribbling on page 0 if that page is accessible (whether mapped or not) to a thread. Thus, if page 0 is accessible, it is important to place it off limits by write- and read-protecting it. This ensures that reads and writes through a NULL pointer result in exceptions. Failing to place page 0 off limits is a sure way to let dangerous faults go undetected until they cause serious problems. It is also good practice to protect a few more low-order pages. The reason for this is that a NULL pointer with a sufficiently large offset (generated by indexing a pointer to an array, for example) could access them.

Ideally, the value of NULL would reference a high-order page, one known to be out of bounds to all threads. However, redefining NULL to a value like `0xfff00000` is dangerous. First, doing so is system specific and therefore leads to interworking problems with third-party software. Second, some programmers will complain about having to write `if(p == NULL)` and `if(p != NULL)` instead of the idiomatic `if(!p)` and `if(p)`. Redefining NULL is therefore inadvisable unless a system uses no third-party software and cannot place page 0 out of bounds for some reason.

7.2.3 Protecting Critical Data

Although trapping references to low-order pages detects users of NULL pointers, it does not detect users of stale pointers. A stale pointer is one that references deleted data. The memory where the data used to reside may subsequently have been reassigned. Yet another problem is using a pointer that has been corrupted by a trampler or some other logic error. Using a stale or corrupt pointer is likely to cause further corruption.

An extreme system must protect its data from trampling by stale or corrupt pointers. The rest of this chapter discusses techniques that serve this purpose.

7.2.4 User Spaces

Many operating systems provide User Spaces, which are one way to protect data from corruption. Each process runs in its own user space, which is a private memory segment. Other memory segments are inaccessible to the process, and any attempt to reference one causes an exception. User spaces place firewalls between processes, thus preventing one process from trampling data in another process. Running processes in user spaces is a common technique in the computing industry.

7.2.5 Disadvantages of User Spaces

Later in this chapter, we will return to user spaces and discuss situations in which it is appropriate to use them in an extreme system. However, they also have drawbacks that prevent them from being a universal solution to the problem of memory corruption. Because the desire to overcome these drawbacks leads to a different solution, we will look at them first:

- User spaces make context switches more expensive because each context switch involves remapping memory.
- User spaces incur further overhead when entering and exiting kernel mode. That is, when an operating system call needs to access memory that is off-limits to the process making the call, it must remap memory at the beginning and end of the call.
- User spaces do not prevent a process from corrupting its *own* data. A process usually contains many threads, any of which could cause such a corruption. If the process or its data is important, an outage is then the likely outcome.
- User spaces arose in timesharing systems, where processes are unrelated and untrustworthy, but the processes in an embedded extreme system collaborate extensively and require thorough testing. When two processes frequently collaborate, putting an absolute firewall between them significantly reduces capacity because they must communicate with messages. Take the simple case of one process wanting to read data owned by another process. If the data is directly accessible, the cost will probably be in microseconds, but if a message must access the data, the cost will probably be measured in milliseconds – about 1000 times more expensive. If this proves unacceptable – and it will, if the messages occur with sufficient frequency – the solution will either be to merge the processes or to move the data of mutual interest to a shared segment, thereby increasing its vulnerability.

This last solution, moving frequently accessed data to a shared segment, is a common solution for avoiding the unacceptable cost of user spaces. The need to prevent corruption in the shared segment leads to the next technique.

7.2.6 Write-Protected Memory

When data resides in a shared segment to avoid the cost of accessing it with interprocess messages, its vulnerability increases

Write-protecting the shared segment addresses this drawback. Applications can still read the data directly, but any application that inadvertently writes to the data's protected segment encounters an exception. To update the data, the application must first unprotect the shared segment explicitly. When the application finishes updating the data, it reprotects the shared segment.

This solution, WRITE-PROTECTED MEMORY, significantly improves capacity while still providing protection against tramplers. However, only data that changes infrequently can be write protected. If the data changes too often, the need constantly to unprotect and reprotect it causes too much processing overhead.

Write-protected memory is often superior to USER SPACES. Only write protection is needed to prevent trampling. Read protection degrades capacity while serving no useful purpose. C++, not the memory management system, should provide encapsulation!

The object code in most systems is write protected. An extreme system usually contains application data that is equally worthy of protection, but what data?

In part, the answer is configuration data, including subscriber profiles and the system's physical configuration, such as what cards are in what slots and what their addresses are. This data is essential to the system's ongoing operation and meets the criterion of being infrequently updated. Because it is updated infrequently, the term **static data** can be used to describe it. This is in contrast to **dynamic data**, which frequently changes and includes things like run-time instances of state machines.

Protecting configuration data safeguards it against tramplers and increases the odds of discovering any tramplers that exist. However, it also confers another important benefit: if static data can be trusted, then a node can return to service much faster if it suffers an outage and reinitializes itself. During such a reinitialization, all dynamic data (essentially, work in progress) is lost. However, the node can quickly begin to process new work again, because it does not have to reload any of its static data. If, for example, it had to reload all of its subscriber profiles, the outage would last much longer. This form of fast reinitialization is part of ESCALATING RESTARTS, which we will discuss later.

Now that we have identified some data to protect, the next question is how to segment the protected data. The easiest approach is simply to use one logical segment, all pages of which are unprotected whenever any of the data needs to change. This approach is reasonably safe because data is only unprotected briefly, and it avoids the complexity that arises if applications have to decide which specific pages to unprotect during a particular update. A configuration parameter determines the size of the write-protected segment, which is created during system initialization. Consideration must also be given to supporting the in-service creation of additional protected segments, which may be needed when expanding the system. Logically, however, the system treats all of the segments as one when disabling or enabling write protection.

Once a system starts to use write-protected data, it soon becomes apparent that the functions `MemUnprotect` and `MemProtect` must be nestable. Assume that there are two functions, both of which update protected data, and both of which therefore surround their updates with calls to `MemUnprotect` and `MemProtect`. Further, assume that one of these functions also calls the other to perform part of its work. When this nesting occurs, two calls to `MemUnprotect` precede two calls to `MemProtect`, so these functions must use a counter.

Yet even this is inadequate if preemptive scheduling can cause a context switch while static data is unprotected. This situation is undesirable because the system will run unprotected for much longer than necessary, until the preempted thread finally gets to complete its work. `MemUnprotect` and `MemProtect` must therefore be member functions of `Thread`, and `unprotectCount_` must be a data member of `Thread`. When a thread is scheduled out, write protection is reenabled. When a thread is scheduled in, write protection is disabled if the thread's `unprotectCount_` is greater than zero. The challenge is making this happen if you can't modify the scheduler. However, you can approximate it.

If preemption is possible, most threads should nonetheless effect COOPERATIVE SCHEDULING by acquiring and releasing the run-to-completion lock. Furthermore, a thread should only call `MemUnprotect` while locked, because its data modifications typically belong in a critical region. This will prevent all other threads that run locked from being scheduled in while write protection is disabled. The functions `Thread::Lock` and `Unlock` can then reenable or disable write protection, based on the status of the threads that are releasing and acquiring the lock. In other words, most context switches occur within these functions, which can therefore perform the write-protection fix-up. Although this is not a complete solution, given that an *unlocked* thread can still be scheduled in while memory is unprotected, it is better than nothing.

Write-protected memory is shared memory, so its users should follow design principles that apply to shared memory. For example, a protected object or data structure should usually have one owner. Although any application may read the data, only the owner may change it. Except for trivial changes, applications that need to change the data should send the owner a message instead of directly invoking its functions. The owner can enforce these restrictions by making some of its functions `public` but others `protected` or `private`.

7.2.7 Protecting Objects

Given that we are implementing software in an OO language (see Chapter 3), objects will own most of the data that we want to protect. It must therefore be possible to place objects in WRITE-PROTECTED MEMORY. In addition to objects that contain subscriber profiles or system configuration data, singletons often merit write protection. A state machine, for example, can be implemented using many singletons. Each state machine consists of a set of states and event handlers. Each state singleton contains an array that is indexed by an event identifier to find the event handler identifier for some state–event combination. Each event handler is also a singleton. Although event handlers could be invoked using function pointers, implementing them as simple objects allows them to be subclassed. All run-time instances of a state machine share its singleton states and event handlers, with each run-time instance containing a reference to its current state. If any of the singleton states or event handlers became corrupted, an exception would occur each time it was used. Consequently, all of them merit write protection.

The question is how to create and destroy objects that reside in write-protected memory. To support this, we introduce the class `ProtectedObject`, which derives from `Object`. It simply overrides operators `new` and `delete` to allocate and free memory in a write-protected segment that comprises a separate heap. A subclass of `PooledObject` can also provide its own versions of operators `new` and `delete` that dequeue and enqueue blocks in an `ObjectPool` created from this heap.

```
void *ProtectedObject::operator new (size_t size)
{
   ProtectedHeap *h =
      SingletonObject<ProtectedHeap>::Instance();
   return h->Allocate(size);
}

void ProtectedObject::operator delete (void *obj)
{
   ProtectedHeap *h =
      SingletonObject<ProtectedHeap>::Instance();
   return h->Free(*((ProtectedObject*) obj));
```

```
}

void ProtectedObject::operator delete (void *msg,
                                       size_t size)
{
   ProtectedHeap *h =
      SingletonObject<ProtectedHeap>::Instance();
   return h->Free(*((ProtectedObject*) obj), size);
}
```

Before code creates or destroys an object that derives from `ProtectedObject`, it must disable write protection. Although operator `new` could disable write protection, this leaves the question of who would eventually reenable it. The leaf constructor could reenable it, but a constructor may not know that it is a leaf. Another option is to disable and reenable write protection in operators `new` and `delete`, as well as in every constructor and destructor – in short, to do so in any function that modifies the object's data. This would make the object's protected status transparent to its users. Unfortunately, it would also waste far too much time unprotecting and reprotecting memory. It is therefore reasonable to insist that applications surround their updates to protected objects with calls to `MemUnprotect` and `MemProtect`.

If a class contains both static and dynamic data, it must be split into two classes, one that places objects in write-protected memory and another that places objects in unprotected memory. A class that represents a card, for example, might place the card's slot and IP address in protected memory but the card's current state (a reference to one of its state singletons) in unprotected memory. This introduces some artificial complexity but helps to achieve carrier-grade quality. Splitting the class is also likely to cause a capacity penalty as the result of collaborating function calls, although this can be alleviated by making the pair of classes friends or by inlining accesses to frequently used data.

7.2.8 When Are User Spaces Appropriate?

Although it is generally preferable to use WRITE-PROTECTED MEMORY instead of USER SPACES, there are situations in which user spaces are a reasonable choice. Say that a system consists of a number of executables that run on different processors. It then makes sense for a small-scale version of the system, running on one processor, to run each executable as a separate process, in its own user space. This avoids the need to create a custom load that combines the executables. It also ensures that they will remain independent, because no one will be able to use shared memory to couple them closely. An example would be combining a call server and media server in a single blade. In a large system, some blades would act as call servers, and others would act as media servers. A small-scale version of the same system might combine these servers on one blade, in separate user spaces.

Applications that are fairly independent at run time can run in their own user spaces. Each telnet or CLI session, for example, might have its own user space.

It might also be reasonable to place subscriber profiles in their own user space. In a telephone switch, this would reduce the system's capacity because both call processing and subscriber administration software need to access these profiles. However, making the profiles inaccessible to call processing offers them some protection against corruption. The call processing software must then fetch a subscriber's profile at the beginning of each call, by sending a message to the subscriber administration software. Afterwards, it runs the call using a copy of the profile. Although this is less efficient than using write-protected memory for profiles, it might be acceptable if the cost of messaging to obtain a profile was a modest percentage of the total cost of processing a call. Whether a subscriber profile resides in a user space or write-protected memory, making a copy of it has the advantage that administration software can then change the original profile while the call is in progress. When a call references the original instance of the profile instead, the profile is usually locked (that is, made read only) for the duration of the call, to prevent inconsistencies in the call's behavior. Using a variant of OBJECT TEMPLATE to block-copy the profile reduces the cost of cloning it at the beginning of the call.

A common claim is that user spaces improve survivability because the loss of one process does not affect other processes. However, this is generally untrue for processes in an extreme system, because most of them are critical. Hence, putting applications in separate processes is not a panacea, because the loss of any process still affects all the processes that collaborate with it. User spaces usually implement an intraprocessor form of HETEROGENEOUS DISTRIBUTION. That is, functions performed on behalf of subscribers run in different processes. The loss of one of these processes therefore causes a ripple effect in which other processes must also clean up their work. To avoid this outcome, a process must checkpoint its objects to a memory segment that will survive if the process dies. The next incarnation of the process can then recreate the objects when it starts to run. However, this increases complexity and the time needed for recovery, so it is not always a viable solution.

What would be truly desirable, from a survivability standpoint, would be to run each *subscriber* in its own user space. All of the work performed on behalf of the subscriber would run in that user space, so that a crash in that software would only affect one subscriber. In this approach, user spaces would provide an intraprocessor form of HOMOGENEOUS DISTRIBUTION. Unfortunately, this would require thousands of processes. Even if the operating system supported that many processes, the result would be an unacceptable loss of capacity.

To conclude, the most compelling case for user spaces is to run different software loads (executables) on the same processor, typically in a cost-reduced configuration. In most other cases, it is preferable to use write-protected memory and run all threads in a single user space, or even in kernel mode, to avoid the overhead of user spaces.

7.3 SUMMARY

- Practice DEFENSIVE CODING. Check pointers, array indices, arguments, and messages. Assume that functions and requests will fail.
- Use STACK OVERFLOW PROTECTION, LOW-ORDER PAGE PROTECTION, AND WRITE-PROTECTED MEMORY to prevent tramplers from corrupting memory.
- USER SPACES increase your addressable market by allowing you to run different executables on the same processor in a cost-reduced configuration.

8

Recovering from Software Faults

Although the techniques discussed in the previous chapter reduce the number of software faults, faults will nonetheless occur. This chapter describes ways to handle errors that arise from them. We will first look at ways to keep a node in service when an error occurs. However, if an error is too severe or the frequency of errors is too high, the best course of action is reinitialization, so we will also look at techniques that apply during those situations.

8.1 SAFETY NET

When a trap occurs, an extreme system must initiate recovery procedures on the affected thread. In this book, the term **trap** refers to anything that can cause the death of a thread. A trap is either a C++ exception (such as `bad_alloc`) or a POSIX signal (such as `SIGSEGV`), which is fatal unless handled.

Here are some examples of traps, roughly in order of most likely to least likely, although the order will differ from one system to the next:

- using an invalid pointer (`SIGSEGV` or `SIGBUS` signal);
- throwing an exception when an application detects a fatal error;
- running locked too long (`SIGVTALRM`, when used as described in Section 5.4.3);
- overflowing the stack (`SIGSEGV`);

Robust Communications Software G. Utas

- being killed (SIGKILL, with SIGCHLD to the parent thread);
- invoking an illegal opcode (SIGILL, the result of a corrupt function pointer);
- dividing by zero (SIGFPE);
- failing to allocate memory in operator new (bad_alloc exception).

This is only a partial list of possible traps. There are many other exceptions and signals, but we need to recover from all of them. A SAFETY NET does this by adding some functions to the Thread class that first appeared in Section 5.6:

- The Start function serves as the common entry function for each Thread. It catches all exceptions and signals to provide a safety net for each thread.
- After Start prepares the safety net, it invokes a virtual Enter function that serves as the entry point to a Thread subclass.
- When Start catches an exception or signal, it invokes a virtual Recover function that allows a Thread subclass to clean up its resources. In most cases, Start then reinvokes the Enter function so that the thread will resume execution without being killed.

The rest of this section describes how Thread provides a safety net and how this allows threads to recover from traps. For the implementation, see Section 8.1.5.

8.1.1 Catching Exceptions

The Start function enters a loop that invokes Enter within a try block. This try block precedes a series of catch blocks that handle all uncaught exceptions. When Start catches an exception, it generates a log and invokes the Recover function so that the thread can release any resources that would otherwise be orphaned. Each Thread subclass must therefore track such resources in its subclass-specific data.

After Start invokes Recover, it reenters its loop to call the thread's Enter function again. However, Start also uses a LEAKY BUCKET COUNTER (see Section 8.2) to track the number of times that a thread has recently trapped. If the number of traps in a defined time interval exceeds a defined threshold, Start kills the thread instead of reentering it.

8.1.2 Catching Signals

Before `Start` invokes a thread's `Enter` function, it registers a common **signal handler** (a static function) against all fatal signals. The signal handler simply throws a `SignalException` whose data specifies the signal that occurred. The `SignalException` shows up in `Start`'s `catch` block, which can then handle the signal in the same way as a standard C++ exception.

Well, at least that's the theory. Although POSIX specifies how to support signals, the details typically vary from one compiler or operating system to the next. To handle signals, you will need to read your compiler and operating system documentation, do some prototyping, and maybe even contact the relevant software supplier for assistance. For example, the GNU C++ compiler contains an option (*-fnon-call-exceptions*) to generate code that allows a signal handler to throw an exception. For further information, see [GNUSIG].

Some compilers or operating systems do not support throwing exceptions from a signal handler. Regrettably, C++ does not mandate this requirement, which testifies to the lack of support for extreme techniques within the computing industry. The real-time Java standard, perhaps because it is more recent, gets this right.

If you cannot manage to turn signals into exceptions, you could fall back to the C functions `setjmp` and `longjmp` [GNUNLE]. Each `Thread` object now requires a `jmp_buf` instance, which is set within `Start` by calling `setjmp`. The signal handler then uses `longjmp` to 'warp' execution back into `Start`, where you can initiate recovery procedures.

The problem with `setjmp` and `longjmp` is their interaction with local objects. A local object is one that must be deleted before a function returns:

- It was constructed on the stack but owns some resource, so it must be deleted to free the resource.
- It was allocated using `new` and is referenced by a local pointer variable, so it must be deleted to free its memory.

Although some C++ compilers unwind the stack properly during a `longjmp`, by deleting local objects, the C++ standard does not mandate this. Using `longjmp` therefore restricts your programming model to one in which functions must not create local objects that require destruction in order to avoid resource leaks. However, this restriction is not too severe because extreme systems allocate local objects far less often than do other systems. The reason is that most of their objects are involved in session processing and must therefore survive from one transaction to the next. These objects are usually owned globally, in hierarchical object models, rather than by individual threads. However, if a function *does* allocate a local object, it must save a reference to the object in its `Thread` subclass, so that the thread's `Recover` function can delete it.

8.1.3 Fatal Errors

Catching exceptions and signals allows a thread to be recovered without being killed. However, there are situations in which this is impossible:

- When a thread is killed explicitly. The `SIGKILL` signal used for this purpose is unconditional: a signal handler cannot catch it. Although killing a thread is rare, it is appropriate when a thread is trapping too often or using too many resources.
- When the signal handler cannot throw an exception. This can occur because of compiler or operating system limitations (Section 8.1.2) or insufficient stack space (Section 7.2.1). However, `setjmp` and `longjmp` can substitute for exceptions, and `sigaltstack` addresses the stack space problem. In the absence of these functions, the signal handler must reraise a signal after generating a debug log. This will cause the thread to be killed.

When a thread is killed, its parent thread receives the `SIGCHLD` signal, which can prompt it to recreate the child thread. As an alternative to `SIGCHLD`, the parent can use the technique described in Section 13.3.5 to detect and recreate a killed child.

Software that handles exceptions and signals must be bulletproof because the occurrence of a second signal or exception, while handling the original one, usually leads to grief. A nested exception causes a crash, and a nested signal may cause an infinite loop that endlessly invokes the signal handler. Trap recovery code therefore requires rigorous code inspection and thorough testing. Such testing involves modifying code to cause traps. It is therefore white box in nature, meaning that it based on the system's internal structure.

8.1.4 Recovering Resources

When a thread is recovered, or killed and recreated, it must release any resources that were orphaned because of the trap. If the thread is a daemon, which is the case for most threads in an extreme system, it should not need to release many resources. The reason for this is that the thread serves many clients. Consequently, it only needs to release the resources associated with the specific work that it was performing when the trap occurred.

Consider an I/O thread that receives all of the messages arriving on some IP port and places these messages on a work queue. After a trap, the thread should not release the socket that receives messages, because it will continue to use the socket when it resumes its execution. If it were to release the socket, it would lose all unread messages, and it would have to allocate a new socket.

However, the I/O thread probably allocates a buffer each time it receives a message from the socket. In order to place this buffer on a work queue, the I/O thread invokes application-specific code that wraps the buffer and identifies the work queue on which to place it. If a trap occurs while this application software is handling the buffer, the I/O thread must release the buffer when it resumes execution. For this purpose, the I/O thread can define a pointer to the buffer in its `Thread` subclass. When the I/O thread allocates a buffer, it sets this pointer. After it invokes the application code and places the buffer (perhaps wrapped) on a work queue, it clears the pointer. Thus, if a trap occurs while processing the buffer, the pointer will be non-NULL when the thread's `Recover` function is invoked, which allows the thread to free the buffer. An alternative is to use an `auto_ptr` to reference the buffer so that it will be released if an exception unwinds the stack. Putting the pointer in the `Thread` subclass is necessary, however, if the signal handler uses `longjmp` as described earlier.

As a second example, consider an invoker thread that dequeues a message from a work queue and passes it to an application's state machine for processing. A trap in the invoker thread, therefore, usually occurs in application software. In this case, the invoker thread's `Recover` function must restrict its cleanup to the specific state machine that was running at the time of the trap. Afterwards, the invoker thread's `Enter` function is reentered, and it continues with the next message on the work queue. By releasing only those objects associated with the transaction that trapped, the invoker thread ensures that other messages on the work queue, and other state machines, remain unaffected. This is another argument for using COOPERATIVE SCHEDULING and state machines. A carefully designed state machine tracks its transient use of resources so that it knows what to clean up when an error occurs.

It is dangerous, however, for the invoker thread's `Recover` function to simply `delete` the objects associated with a transaction that trapped. Quite possibly, the reason for the trap was that one of these objects was in an invalid state. If this is the case, using `delete` might result in another trap – and therefore an outage – because of a trap while handling a trap. A better approach is simply to mark the affected objects for cleanup and `delete` them later, when a second trap will not prove fatal. Finally, to prevent the objects from causing recurring traps, they should be marked as corrupt before they undergo deletion. Then, if they actually do cause another trap, the software that cleans them up will notice that it already tried to do so once, but failed. Consequently, it can limit the second cleanup attempt to simply freeing the memory occupied by them.

8.1.5 Implementing the `Thread` Class

Now that we have discussed all of the extreme techniques that involve threads, let's look at the details of the `Thread` class. `Thread` uses a lock to implement run-to-completion and `SIGVTALRM` to enforce the RUN-TO-COMPLETION TIMEOUT. Its signal handler can either invoke `longjmp` or throw an exception.

`Thread` supports COOPERATIVE SCHEDULING fairly efficiently, but one inefficiency should be noted: a blocking operation can result in *two* context switches instead of one. The first one can occur when `EnterBlockingOperation` releases the run-to-completion lock. If a higher priority thread is waiting for the lock, it runs immediately. When the original thread resumes execution, it actually performs the blocking operation, which can result in a second context switch. In practice, however, this inefficiency should rarely occur. Under PROPORTIONAL SCHEDULING, priorities do not exist. And in the absence of proportional scheduling, almost all of the system's threads should run at the same priority, as discussed in Section 5.7.3.

`Thread` uses the standard versions of the functions `signal`, `raise`, `setjmp`, and `longjmp`. It also assumes that the following abstraction layer can be targeted to the underlying operating system:

```
typedef void* lock;
const int forever = -1;         // for sleep()
const int ITIMER_VIRTUAL = 1; // for setitimer()

void create_thread(void (*start)(void *arg),
                   void *arg, int faction,
                   int stacksize);
int  getpid(void);
void sleep(int msecs);
void interrupt(int pid);
void clear_interrupt(void);
```

```
void kill(int pid);
uint ticks_now(void);
uint secs_to_ticks(uint secs);
void setitimer(int timer, int msecs);
lock create_lock(void);
void acquire_lock(lock l);
void release_lock(lock l);
```

All `Thread` instances are placed in a global `ThreadRegistry`. `Thread` also defines a few subclasses of `exception`.

```
typedef int ThreadId; // identifier assigned by
                      // ThreadRegistry
const ThreadId NilThreadId = -1;
const ThreadId MaxThreadId = 255;

enum ThreadFaction // for proportional scheduling
{
   IdleFaction,
   AuditFaction,
   BackgroundFaction,
   ProvisioningFaction,
   MaintenanceFaction,
   PayloadFaction,
   PriorityFaction
};

enum ThreadRc // what to do after a thread traps
{
   DeleteThread,  // exit the thread
   ReenterThread, // reinvoke the thread's Enter
                  // function
   ForceRestart   // cause a restart
};

class Thread: public Object
{
   friend class ThreadRegistry;
public:
   virtual ~Thread(void);
   static void   MutexOn(void);
   static void   MutexOff(void);
   static int    MsecsLeft(void);
   static void   Pause(int msecs);
   void          Interrupt(void);
   static bool   EnterBlockingOperation(void);
   static void   ExitBlockingOperation(void);
   static void   MemUnprotect(void);
```

```
   static void   MemProtect(void);
   static void   CauseTrap(void);
   static void   CauseRestart(int code, int value);
   ThreadFaction Faction(void) { return faction_; };
   ThreadId      Tid(void) { return tid_; };
   bool          Blocked(void) { return blocked_; };
   void          Start(void); // provides safety net
   static void   CaptureStack(ostringstream &stack,
                             bool full);
protected:
   Thread(ThreadFaction faction, int stacksize);
   //
   // The second constructor wraps the primordial
   // thread. Subclasses provide Enter and Recover.
   //
   Thread(int pid, ThreadFaction faction);
   virtual void     Enter(void) = 0;
   virtual ThreadRc Recover(void) = 0;
   virtual bool     BlockingAllowed(void);
   virtual void     LogTrap(exception *ex,
                            ostringstream &log);
private:
   // EnterThread is the true entry function.
   // SignalHandler catches signals. Unprotect and
   // Reprotect support write-protected memory.
   // Lock prevents a thread from being scheduled
   // out until it calls Unlock.
   //
   static void EnterThread(void *arg);
   static void SignalHandler(int sig);
   void RegisterForSignal(int sig);
   void RegisterForSignals(void);
   static void Unprotect(void);
   static void Protect(void);
   void Lock(void);
   void Unlock(void);
   bool HandleTrap(exception *ex, int sig,
                   const string &stack);
   void ResetTrapLog(bool realloc);

   static Thread *LockedThread;      // thread running
                                     // locked
   static lock RunToCompletionLock; // global lock
   static bool SafetyNetOn;          // enables signal
                                     // handler
   static bool LongjmpOnSignal;      // controls signal
                                     // handler
```

```
   ThreadFaction faction_;     // scheduler faction
   int pid_;                   // id from getpid()
   ThreadId tid_;     // assigned by ThreadRegistry
   bool entered_;     // true if execution has started
   bool recovering_; // true if recovering from a trap
   bool kill_;                 // true if to be killed
   uint startTime_;            // time when starting to
                               // run locked
   bool blocked_;              // true if blocked
   bool interrupted_;          // true if interrupted
   int mutexCount_;            // >0 if running locked
   int unprotectCount_;        // >0 if writing to
                               // protected memory
   int trapCount_;             // number of traps
   jmp_buf jmpBuffer_;         // for setjmp/longjmp
   ostringstream *trapLog_;    // to build trap log
   ostringstream *stackLog_; // to capture stack trace
};

class ThreadRegistry: public Object
{
   friend class SingletonObject<ThreadRegistry>;
public:
   ThreadId AddThread(Thread &thread);
   ThreadId SetRootThread(Thread &thread);
   Thread   *RootThread(void);
   void     RemoveThread(Thread &thread);
   static Thread *RunningThread(void);
protected:
   ThreadRegistry(void);
   ~ThreadRegistry(void);
private:
   Thread   *threads_[MaxThreadId + 1]; // registry
                                        // of threads
   ThreadId root_; // primordial thread
};

class ExceptionWithStack: public exception
{
public:
   virtual ~ExceptionWithStack() { };
   string stack(void) const { return stack_; };
protected:
   ExceptionWithStack(void);
private:
   string stack_; // contains stack trace
};
```

```
class FatalException: public ExceptionWithStack
{
public:
   FatalException(int code, int value);
   ~FatalException() { };
   int code(void)  const { return code_; };
   int value(void) const { return value_; };
   const char *what(void)
                   const { return "fatal error"; };
private:
   int code_;  // from Thread::CauseRestart argument
   int value_; // from Thread::CauseRestart argument
};

class SignalException: public ExceptionWithStack
{
public:
   SignalException(int sig);
   ~SignalException() { };
   int signal (void) const { return signal_; };
   const string reason(void) const;
   const char *what(void) const;
   static const string signalString(int sig);
private:
   int signal_; // the actual signal (e.g. SIGSEGV)
};
```

And now the implementation:

```
lock Thread::RunToCompletionLock =  create_lock();
Thread *Thread::LockedThread = NULL;
bool Thread::SafetyNetOn = true;
bool Thread::LongjmpOnSignal = false;

const int RunToCompletionTimeout = 20;  // msecs
const int DeathOfCriticalThread = 1;
const string strUnknownException =
               "unknown exception";
const string strStackFrame = "Stack Frame";
const string strStackUnavailable = "unavailable";
const string strTrapDuringRecovery =
               "Trap during recovery.";

void Thread::Lock(void)
{
   // Acquire the run-to-completion lock. Record the
   // locked thread and the time when it started
```

```
    // running. Start the run-to-completion timer
    // and unprotect memory if necessary.
    //
    if(LockedThread == this) return;
    acquire_lock(RunToCompletionLock);
    LockedThread = this;
    startTime_ = ticks_now();
    setitimer(ITIMER_VIRTUAL, RunToCompletionTimeout);
    if(unprotectCount_ > 0) Unprotect();
}

void Thread::Unlock(void)
{
    // Reprotect memory, cancel the run-to-completion
    // timer, clear the locked thread, and release the
    // run-to-completion lock.
    //
    if(LockedThread != this) return;
    if(unprotectCount_ > 0) Protect();
    setitimer(ITIMER_VIRTUAL, 0);
    LockedThread = NULL;
    release_lock(RunToCompletionLock);
}

void Thread::Unprotect(void)
{
    ProtectedHeap *heap =
        SingletonObject<ProtectedHeap>::Instance();
    heap->Unprotect();
}

void Thread::Protect(void)
{
    ProtectedHeap *heap =
        SingletonObject<ProtectedHeap>::Instance();
    heap>Protect();
}

void Thread::MutexOn(void)
{
    // This must be nestable. If the thread is not
    // already locked, schedule it out before it
    // begins to run locked so that it can only run
    // for as long as the run-to-completion timeout.
    //
    Thread *thread = ThreadRegistry::RunningThread();
    thread->mutexCount_++;
```

```
   if(thread->mutexCount_ == 1) Pause(0);
}

void Thread::MutexOff(void)
{
   // If the thread no longer needs to run locked,
   // schedule it out so that other threads can run
   // first.
   //
   Thread *thread = ThreadRegistry::RunningThread();
   if(thread->mutexCount_ <= 0)
   {
      thread->mutexCount_ = 0;
      return;
   }
   thread->mutexCount_--;
   if(thread->mutexCount_ == 0) Pause(0);
}

int Thread::MsecsLeft(void)
{
   // Tell the thread how much longer it can run.
   // This could also be implemented with getitimer,
   // but this is inaccurate on many systems.
   // However, the implementation below is also
   // inaccurate if the scheduler can preempt
   // a locked thread to run an unlocked thread.
   // Should this occur, we will understate how
   // much time is left.
   //
   Thread *thread = ThreadRegistry::RunningThread();
   return (RunToCompletionTimeout -
            (ticks_now() - thread->startTime_));
}

void Thread::Pause(int msecs)
{
   // If the thread has been interrupted, it has more
   // work to do, so it should only sleep for "zero
   // time". If msecs is negative, the thread wants
   // to sleep "forever" (until interrupted).
   //
   Thread *thread = ThreadRegistry::RunningThread();
   if(thread->interrupted_)
   {
      thread->interrupted_ = false;
      msecs = 0;
   }
```

```
   else if(msecs < 0) msecs = forever;

   EnterBlockingOperation();
   sleep(msecs);
   clear_interrupt();  // in case we were interrupted
   ExitBlockingOperation();
}

void Thread::Interrupt(void)
{
   // If the thread is blocked, wake it up.
   //
   if(blocked_)
      interrupt(pid_);
   else
      interrupted_ = true;
}

bool Thread::BlockingAllowed(void)
{
   return true;  // can be overridden by subclasses
}

bool Thread::EnterBlockingOperation(void)
{
   // When a thread could block, it must release
   // the lock.
   //
   Thread *thread = ThreadRegistry::RunningThread();
   if(!thread->BlockingAllowed()) return false;
   if(LockedThread == thread) thread->Unlock();
   thread->blocked_ = true;
   return true;
}

void Thread::ExitBlockingOperation(void)
{
   // When a previously locked thread resumes running
   // after using a blocking operation, it becomes
   // locked again.
   //
   Thread *thread = ThreadRegistry::RunningThread();
   thread->blocked_ = false;
   if(thread->mutexCount_ > 0) thread->Lock();
}

void Thread::MemUnprotect(void)
```

```
{
   Thread *thread = ThreadRegistry::RunningThread();
   if(thread->unprotectCount_ == 0)
   {
      MutexOn();  // update write-protected data in a
                  // critical region
      Unprotect();
   }
   thread->unprotectCount_++;
}

void Thread::MemProtect(void)
{
   Thread *thread = ThreadRegistry::RunningThread();
   thread->unprotectCount_--;
   if(thread->unprotectCount_ == 0)
   {
      Protect();
      MutexOff();  // undo MutexOn in MemUnprotect
   }
   if(thread->unprotectCount_ < 0)
      thread->unprotectCount_ = 0;
}

Thread::Thread(ThreadFaction faction, int stacksize)
{
   faction_         = faction;
   pid_             = -1;
   tid_             = -1;
   entered_         = false;
   recovering_      = false;
   kill_            = false;
   startTime_       = 0;
   blocked_         = false;
   interrupted_     = false;
   mutexCount_      = 0;
   unprotectCount_  = 0;
   trapCount_       = 0;
   trapLog_         = NULL;
   stackLog_        = NULL;

   // Create the thread and add it to the registry.
   //
   create_thread
      (
         EnterThread, // entry function for thread
         this,        // argument will be this object
```

```
         faction,     // thread's faction
         stacksize    // size of thread's stack
      );
   ThreadRegistry *reg =
      SingletonObject<ThreadRegistry>::Instance();
   tid_ = reg->AddThread(*this);
}

Thread::Thread(int pid, ThreadFaction faction)
{
   // The primordial thread calls this to get wrapped.
   //
   faction_         = faction;
   pid_             = getpid();
   tid_             = -1;
   entered_         = true;
   recovering_      = false;
   kill_            = false;
   startTime_       = ticks_now();
   blocked_         = false;
   interrupted_     = false;
   mutexCount_      = 0;
   unprotectCount_  = 0;
   trapCount_       = 0;
   trapLog_         = NULL;
   stackLog_        = NULL;
   ThreadRegistry *reg =
      SingletonObject<ThreadRegistry>::Instance();
   tid_ = reg->SetRootThread(*this);
}

Thread::~Thread(void)
{
   // If the thread was locked, make sure that things
   // get cleaned up. When a thread is killed, the
   // O/S must release any lock that the thread
   // holds. If it didn't, we would have to allocate
   // a new run-to-completion lock to prevent the
   // system from hanging.
   //
   if(LockedThread == this) LockedThread = NULL;
   ThreadRegistry *reg =
      SingletonObject<ThreadRegistry>::Instance();
   reg->RemoveThread(*this);
   if(pid_ > 0) kill(pid_);
}
```

```
void Thread::EnterThread(void *arg)
{
   // ARG actually references a Thread object. When
   // the Thread constructor creates a thread, the
   // thread may start to run immediately, before
   // the Thread object is fully constructed. This
   // leads to trouble (invocation of a pure virtual
   // Start function). Sleep for 2 seconds to give
   // the thread time to be constructed.
   //
   sleep(2000);
   ((Thread*) arg)->Start();
}

void Thread::SignalHandler(int sig)
{
   if(!SafetyNetOn)
   {
      // Turning off the safety net in the lab allows
      // the debugger to be entered. Restore the
      // default handling for this signal and reraise
      // it.
      //
      signal (sig, SIG_DFL);
      raise(sig);
      return;
   }
   if(LongjmpOnSignal)
   {
      // This platform does not support throwing
      // exceptions from a signal handler. Use
      // longjmp to return to Thread::Start.
      // Execution resumes at the point where
      // setjmp was invoked.
      //
      Thread *thread = ThreadRegistry::RunningThread();
      thread->ResetTrapLog(true);
      thread->CaptureStack(*thread->stackLog_, true);
      longjmp(thread->jmpBuffer_, sig);
   }
   else
   {
      // Turn the signal into a C++ exception.
      // Thread::Start will catch it and initiate
      // recovery action.
      //
```

```
      throw SignalException(sig);
   }
}

void Thread::RegisterForSignal(int sig);
{
   // This should also call sigaltstack, but the
   // simple version is
   //
   signal (sig, SignalHandler);
}

void Thread::RegisterForSignals(void)
{
   // The signals to register for are system specific,
   // but this list is typical. We do not handle
   // SIGCHLD because the details are quite platform
   // specific. Furthermore, if the safety net works
   // properly, a child should always be able to
   // continue unless it is killed explicitly (by
   // kill, in the destructor) or traps when
   // recovering_ is set.
   //
   RegisterForSignal(SIGABRT);
   RegisterForSignal(SIGBUS);
   RegisterForSignal(SIGFPE);
   RegisterForSignal(SIGILL);
   RegisterForSignal(SIGINT);
   RegisterForSignal(SIGQUIT);
   RegisterForSignal(SIGSEGV);
   RegisterForSignal(SIGSYS);
   RegisterForSignal(SIGTERM);
   RegisterForSignal(SIGVTALRM);
   RegisterForSignal(SIGXCPU);
}

void Thread::LogTrap(exception *ex,
                     ostringstream &log)
{
   // Subclasses override this to add their own
   // data to trap logs.
}

void Thread::CaptureStack(ostringstream &stack,
                          bool full)
{
```

```
   // Capture the function call chain in STACK.
   // If FULL is set, arguments and locals should
   // also be captured. This has to invoke some
   // black magic, platform-specific function.
}

bool Thread::HandleTrap(exception *ex, int sig,
                        const string &stack)
{
   // Build a trap log.
   //
   ResetTrapLog(true);
   *trapLog_ << "EXCEPTION in ";
   *trapLog_ << Object::StrObj(this, false) << endl;
   *trapLog_ << "  type: ";
   if(ex != NULL)
      *trapLog_ << ex->what();
   else if(sig >= 0)
      *trapLog_ << SignalException::signalString(sig);
   else
      *trapLog_ << strUnknownException;
   *trapLog_ << endl;

   // If the thread trapped during recovery, it dies.
   //
   if(recovering_)
   {
      kill_ = true;
      *trapLog_ << strTrapDuringRecovery << "  ";
   }
   LogTrap(ex, *trapLog_);
   *trapLog_ << strStackFrame << endl;
   *trapLog_ << stack << endl;
   send trapLog_ to the log system;
   ResetTrapLog(false);
   return SafetyNetOn;
}

void Thread::ResetTrapLog(bool realloc)
{
   delete trapLog_;
   trapLog_ = NULL;
   if(realloc) trapLog_ = new ostringstream;
}

void Thread::CauseTrap(void)
{
```

```
   // This is useful for testing trap handling.
   //
   int *p = (int*) 0xeeeeeeef;
   if(*p == 0) p++;  // dereferencing P causes
                     // a signal
}

void Thread::CauseRestart(int code, int value)
{
   throw FatalException(code, value);
}

void Thread::Start(void)
{
   for(trapCount_ = 0; true; trapCount_++)
   {
      try
      {
         int       sig;
         ThreadRc trc;

         // Save the thread's identifier. Mark this
         // location so that the signal handler can
         // longjmp to it. Create the safety net for
         // signals.
         //
         pid_ = getpid();
         sig = setjmp(jmpBuffer_);
         RegisterForSignals();

         // If SIG is zero, we just invoked setjmp.
         // If it is non-zero, we got here because
         // the signal handler invoked longjmp.
         //
         if(sig != 0)
         {
            if(HandleTrap(NULL, sig,
                              stackLog_->str()))
               continue;
            return;
         }

         if(trapCount_ > 0)
         {
            // The thread is being reentered after
            // a trap. Tell it to recover and then
            // continue as directed.
```

```
         //
         if(!recovering_)
         {
            recovering_ = true;
            trc = Recover();
            recovering_ = false;
         }

         if((kill_) && (trc == ReenterThread))
            trc = DeleteThread;

         switch(trc)
         {
         case DeleteThread:  return;
         case ReenterThread: break;
         default:
            CauseRestart(DeathOfCriticalThread,
                         pid_);
         }

         // If the thread was running locked,
         // unlock it and force it to yield.
         // Its Enter function will relock it.
         //
         if(mutexCount_ > 0)
         {
            mutexCount_ = 0;
            Pause(0);
         }
      }

      // Record the time when the thread started to
      // run and invoke its entry function.
      //
      startTime_ = ticks_now();
      entered_ = true;
      Enter();
   }

   catch (FatalException &ex)
   {
      ResetTrapLog(true);
      *trapLog_ << "FATAL ERROR in "<< endl;
      *trapLog_ << Object::StrObj(this, false);
      *trapLog_ << endl;
```

```
         *trapLog_ << "  code:  " << ex.code() << endl;
         *trapLog_ << "  value: " << ex.value() << endl;
         *trapLog_ << strStackFrame << endl;
         *trapLog_ << ex.stack() << endl;
         save trapLog_ in the flight recorder;
         ResetTrapLog(false);
         interrupt the primordial thread to cause
            a restart;
      }

      // We can recover from other exceptions.
      //
      catch (ExceptionWithStack &ex)
      {
         if(HandleTrap(&ex, -1, ex.stack())) continue;
         throw(ex);
      }

      catch (exception &ex)
      {
         if(HandleTrap(&ex, -1, strStackUnavailable))
           continue;
         throw(ex);
      }

      catch (...)
      {
         if(HandleTrap(NULL, -1, strStackUnavailable))
            continue;
         throw;
      }
   }
}

ThreadRegistry::ThreadRegistry(void)
{
   int i;
   for(i = 0; i <= MaxThreadId; i++)
      threads_[i] = NULL;
   root_ = NilThreadId;
}

ThreadRegistry::~ThreadRegistry(void)
{
   // Delete threads in reverse order of their
```

```
   // creation.
   //
   int i;
   for(i = MaxThreadId; i >= 0; i--)
   {
      if(threads_[i] != NULL)
      {
         delete threads_[i];
         threads_[i] = NULL;
      }
   }
}

ThreadId ThreadRegistry::AddThread(Thread &thread)
{
   int i;
   for(i = 0; i <= MaxThreadId; i++)
   {
      if(threads_[i] == NULL)
      {
         threads_[i] = &thread;
         return i;
      }
   }
   return NilThreadId;
}

void ThreadRegistry::RemoveThread(Thread &thread)
{
   ThreadId tid = thread.tid_;
   if((tid < 0) || (tid > MaxThreadId)) return;
   if(threads_[tid] != &thread) return;
   threads_[tid] = NULL;
}

ThreadId ThreadRegistry::SetRootThread(Thread &root)
{
   // ROOT has been wrapped. If there is no root
   // thread, assume that it is the root thread,
   // else assume it is a child thread.
   //
   if(root_ == NilThreadId)
   {
      root_ = AddThread(root);
      return root_;
   }
```

```
   return AddThread(root);
}

Thread *ThreadRegistry::RootThread(void)
{
    if(root_ == NilThreadId) return NULL;
    return threads_[root_];
}

Thread *ThreadRegistry::RunningThread(void)
{
   // Most threads run to completion, so the running
   // thread is usually the locked thread. If it
   // isn't, we have to search for it the hard way.
   //
   int pid = getpid();
   Thread *thread = Thread::LockedThread;
   if((thread != NULL) && (thread->pid_ == pid))
      return thread;
   ThreadRegistry *reg =
      SingletonObject<ThreadRegistry>::Instance();
   ThreadId tid;
   for(tid = 0; tid < MaxThreadId; tid++)
   {
      thread = reg->threads [tid];
      if((thread != NULL) && (thread->pid_ == pid))
         return thread;
   }
   return NULL;
}

ExceptionWithStack::ExceptionWithStack(): exception()
{
   ostringstream stack;
   Thread::CaptureStack(stack, true);
   stack_ = stack.str();
}

FatalException::FatalException(int code, int value):
   ExceptionWithStack()
{
   code_  = code;
   value_ = value;
}

const string SigAbrtStr   = "program aborted";
```

```
const string SigBusStr    = "bus error";
const string SigFpeStr    = "floating point error";
const string SigIllStr    = "illegal instruction";
const string SigIntStr    = "program interrupted";
const string SigQuitStr   = "program quit";
const string SigSegvStr   = "segmentation violation";
const string SigSysStr    = "invalid system call";
const string SigTermStr   = "program terminated";
const string SigVtAlrmStr = "virtual alarm timeout";
const string SigXcpuStr   = "program ran too long";
const string SigUnknStr   = "unknown signal";

SignalException::SignalException(int sig):
   ExceptionWithStack()
{
   signal_ = sig;
}

const string SignalException::reason(void) const
{
   return signalString(signal_);
}

const string SignalException::signalString(int sig)
{
    switch(sig)
    {
    case SIGABRT:   return SigAbrtStr;
    case SIGBUS:    return SigBusStr;
    case SIGFPE:    return SigFpeStr;
    case SIGILL:    return SigIllStr;
    case SIGINT:    return SigIntStr;
    case SIGQUIT:   return SigQuitStr;
    case SIGSEGV:   return SigSegvStr;
    case SIGSYS:    return SigSysStr;
    case SIGTERM:   return SigTermStr;
    case SIGVTALRM: return SigVtAlrmStr;
    case SIGXCPU:   return SigXcpuStr;
    }
    return SigUnknStr;
}

const char *SignalException::what(void) const
{
   string s = reason();
   return s.c_str();
}
```

8.2 LEAKY BUCKET COUNTER

A LEAKY BUCKET COUNTER [GAM96] determines when some event has occurred at some threshold frequency. Extreme systems often use leaky bucket counters to track faults that can occur intermittently and infrequently without requiring corrective action. However, if the faults occur with sufficient frequency, the system must react to them.

A leaky bucket counter is initialized with a threshold number of events and the length of the interval during which this many events must occur to empty the bucket. The bucket is notified each time an event occurs. If the bucket is full, an event starts a new interval and decrements the number of events that are still allowed during this interval. Subsequent events continue to drain the bucket. The bucket eventually empties or refills itself once the interval has passed. Here is a sketch of the implementation, assuming (for simplicity) that a timestamp is immune to wraparound:

```
typedef unsigned int   uint;
typedef unsigned short ushort;

class LeakyBucketCounter: public Object
{
public:
   LeakyBucketCounter(ushort events, ushort seconds);
   virtual ~LeakyBucketCounter(void) { };
   bool ThresholdExceeded(void);
private:
   ushort fullSize_;    // number of credits
                        // in full bucket
   uint   fillTime_;    // time required to refill
                        // bucket
   ushort counter_;     // current number of credits
                        // in bucket
   uint   startTime_;   // time when bucket started
                        // to drain
};

LeakyBucketCounter::LeakyBucketCounter
   (ushort events, ushort seconds)
{
   fullSize_  = events;
   fillTime_  = secs_to_ticks(seconds);
   counter_   = events;
   startTime_ = ticks_now();
}
```

```
bool LeakyBucketCounter::ThresholdExceeded(void)
{
   uint currTime = ticks_now();
   if(counter_ == fullSize_)
      startTime_ = currTime;   // bucket starting
                               // to drain now
   else
   {
      if((currTime - startTime_) >= fillTime_)
      {
         // It's time to refill the bucket.
         counter_ = fullSize_;
         startTime_ = currTime;
      }
   }
   counter_--;
   if(counter_ == 0)
   {
      // Refill the empty bucket.
      counter_ = fullSize_;
      startTime_ = currTime;
      return true;
   }
   return false;
}
```

Leaky bucket counters occur in many settings. A common usage is to detect link errors. Occasionally, a packet or message arriving on a link might contain a checksum error because of a transient glitch, but if many errors occur in a short time, the link is probably faulty, so the system should alert the craftspeople by raising an alarm. The system might also autonomously take the link out of service and subject it to an automated diagnostic, after which the link either returns to service (if the diagnostic does not find a fault) or remains offline.

A leaky bucket counter can also monitor a thread's health. If a thread traps infrequently, it is reasonable to recover it, but if it starts to trap regularly, it should be killed. Its parent might then decide to recreate it, although this is unlikely to resolve the problem in the absence of any other corrective action, such as fixing a faulty data structure. Consequently, a leaky bucket counter might also track the total number of processor traps and force a reinitialization if it reaches zero.

A parent thread can also use a leaky bucket counter for each of its child threads. These counters detect death loops (frequently recurring traps) in recreated threads, thereby limiting the number of times that they are recreated. It is a good idea to combine this approach with a back-off scheme, in which the parent waits for successively longer periods before recreating a child that appears to be in a death loop. Eventually the parent might simply stop recreating the child. However, threads in an extreme system are rarely that dispensable: drawing a cute picture of a bomb is not an option.

8.3 AUDIT

In a system that runs continuously, some software faults cause inconsistencies instead of immediately detectable errors such as traps. These inconsistencies include

- orphaned resources;
- incorrect pointers or indices in data, and
- inconsistent states in data.

Eventually these inconsistencies lead to trouble. Orphaned resources slowly degrade throughput and eventually necessitate a reinitialization. Faulty pointers or indices eventually cause traps. Inconsistent states eventually cause behavioral errors.

When a software fault leads to a trap, SAFETY NET deals with the problem reactively, but when a fault leads to inconsistencies, a proactive solution is needed. This is the purpose of an AUDIT.

An extreme system contains a number of audits, each of which focuses on a different part of the system. An audit is a background thread that looks for inconsistencies, logs them, and typically tries to correct them before they lead to trouble. Even if an audit only logs errors instead of correcting them, it highlights hidden faults that designers need to find and fix.

An audit that corrects errors helps to make a system self-healing. However, such an audit must be carefully designed. Even the data structures to be audited require careful design, because some structures are easier to audit than others.

An audit must consider transient states which, on the surface, appear to be inconsistencies but are actually false positives. When an audit inspects data for inconsistencies, it cannot lock the data for too long, as this could prevent payload work from executing. Consequently, an audit might not flag an error or try to correct it unless the audit observes the error twice in succession. A well-designed audit must consider critical regions and keep them as granular as possible. The need for such granularity may, in turn, affect the design of the data structures to be audited, particularly with regard to how they are updated. Data structures that localize changes are easier to audit than those in which updates have a wide impact.

When an audit corrects errors, it must not introduce further inconsistencies. Error correction is difficult because an audit could encounter a wide range of errors. Because it is difficult to anticipate all of them in advance, error correction is often added to an audit in hindsight, in reaction to some type of inconsistency that was actually observed in a running system. The ability to correct errors typically requires some redundant information so that the audit can perform crosschecks. Repairing a two-way queue, for example, might be easier than repairing a one-way queue. Again, the need to audit a data structure may affect its design.

The rest of this section discusses the design of resource pool audits, which are the most common form of audit in extreme systems.

8.3.1 Resource Pool Audit

Each resource pool should provide an audit to recover resources which belong to the pool but that have become orphaned. The algorithm for such an audit typically uses a mark-and-sweep strategy that resembles the following:

1. Mark all resources as orphaned. For this purpose, each object that represents a resource defines an `orphaned` flag. The objects must reside in an array, or some similar data structure, which allows the audit to iterate over all of them, whether or not they are currently in use.
2. Clear the `orphaned` flag for all resources that are not in use. The resource pool typically maintains a free queue of available resources so that it can allocate them quickly, in which case this step simply involves traversing the free queue to claim the idle resources.
3. Invoke application-specific code to claim the resources that are actually in use. When an application owns a resource, it must therefore register it in one of its objects, and it must have a way to iterate through all of its objects to find the resources. Different applications might use the same type of resource. The resource pool must therefore allow multiple applications to register

callback functions. The audit invokes these callbacks during this step, prompting the applications to claim their resources.

4. At this point, all resources (either on the free queue or in use by applications) should have been claimed. Therefore, iterate over all resources in the pool again, recovering those that are still marked orphaned by returning them to the free queue. Generate a log for each recovered resource.
5. Return to step 1: steps 1 through 4 reside in an endless `while (true)` loop.

This algorithm can audit the blocks that implement an OBJECT POOL (see Section 4.3.1). On the surface, the algorithm seems straightforward, yet there are a number of issues to consider. The following sections discuss them.

8.3.2 Critical Regions

The audit contains a number of critical regions, so it must run locked. However, if its resource pool is large, an audit cannot perform all of its steps in a single pass without using too much CPU time. Consequently, it must periodically schedule itself out. It can probably perform steps 1 and 2 while locked, but it may have to schedule itself out after steps 2, 3, and 4. In some cases, step 3 might itself be broken up, with the audit scheduling itself out after invoking each application callback.

When the audit is scheduled out, applications continue to allocate and deallocate resources. Therefore, a resource's `orphaned` flag must be cleared whenever it is allocated or deallocated. If this were not done, and an application were to return a resource to the free queue between steps 2 and 3, the resource would still appear to be orphaned in step 4.

8.3.3 Queue Corruption

Step 2 is perilous. Queues are wonderful – until they are corrupted, when they cause havoc in the form of frequent traps. Step 2 must therefore be able to detect and correct corruptions in the free queue. For this purpose, each resource, as well as the queue header itself, requires a `corrupt` flag. The audit sets this flag before it dereferences the 'next' pointer associated with the queue head or a resource. The dereferencing occurs when the audit invokes the `Claim` function to clear a resource's `orphaned` flag. After invoking this function, the audit clears the `corrupt` flag in the previous resource (or the queue header).

If the audit traps as the result of a corrupt queue link, Safety Net reinvokes the audit thread's `Enter` function, which contains the audit loop. The audit must therefore remember what step it was performing, in this case step 2, so that it can begin to traverse the free queue over again. However, before it dereferences a pointer, it checks the `corrupt` flag associated with the resource (or queue header) in question. If this flag is set, the audit knows that it has encountered a corrupt link, so it truncates the queue at that point. The orphaned, idle resources eventually return to the free queue in step 4.

In an extreme situation, truncation leaves very few resources on the free queue. For example, a corrupt queue header leaves the free queue empty. In this situation, the audit might decide to run locked until the completion of step 4, so that applications do not encounter a flurry of 'no resource available' conditions, something that would be tantamount to a partial outage. Although applications must handle resource allocation failures gracefully, these failures should only occur when a node is running at peak load. A free queue corruption is something else, and so it may be desirable, particularly in the case of a critical resource pool, to rebuild the free queue as fast as possible when a truncation orphans most of the idle resources.

When an audit recovers an orphaned resource and returns it to the free queue, it should generate a log. However, it should not generate a log when it returns an *idle* resource to the queue. The reason is that, when the audit rebuilds a free queue, it recovers many idle resources, and it should avoid causing a flurry of logs. In this situation, the audit should instead limit itself to two logs. The first one states how many resources were on the free queue and how many actually remained after truncation. Later, the second log simply states how many idle resources were recovered and returned to the queue.

Although Safety Net ensures that an audit will survive traps, it is nonetheless important to avoid them. Exception handling is expensive, and a system producing trap logs causes consternation and embarrassment. To avoid trapping, an audit may choose to perform a sanity check before dereferencing a pointer. If the resources in the pool all reside in a single array, for example, then the audit can take a candidate pointer, `p`, and do the following:

```
if((p < &MyPool[0]) || (p > &MyPool[lastIndex]))
   corrupt = true;
else
```

```
{
   offset = (int) p - (int) &MyPool[0];
   corrupt = ((offset % sizeof(MyResource)) != 0);
};
```

In other words, p must be reference a location within the array, and its offset within the array must be an exact multiple of the size of each resource object.

These techniques, for auditing the sanity of queues, apply to queues in general, even those that do not correspond to a set of available resources. All queues in the system should be periodically verified by audits that detect, and possibly try to fix, corruptions. If a critical queue, such as a work queue, becomes corrupted, and its audit cannot repair it, the audit can at least force an autonomous reinitialization. This is generally preferable to letting the node continue to operate, suffering endless traps until someone eventually realizes that there is a serious problem and finds the courage to perform a manual reinitialization.

8.3.4 Traps in Callbacks

Step 3 is even more perilous than step 2. In step 3, the audit invokes callbacks so that each application can claim the resources that it is using. If a callback traps, some in-use resources remain unclaimed. The audit will therefore recover them, which will force the applications that owned them to abort their work.

Because the callbacks that claim resources are application specific, it is difficult to recommend specific techniques to reduce the risk of traps within them. However, some guidelines merit consideration.

First, if a callback maintains state information and uses a `corrupt` flag, in much the same way as the audit does when traversing its free queue, then the audit can reinvoke the callback, which can then skip any corrupt resources.

Second, the more a callback needs to traverse objects in order to arrive finally at the resources that it claims, the greater the risk that it will encounter a faulty pointer that leads to a trap. To reduce this risk, each resource can invoke an application callback individually. When an application allocates a resource, it registers two items: its callback (a singleton that wraps a `Claim` function) and an `Object` (the resource owner). When the audit invokes the callback, the callback casts the `Object` pointer to the type of subclass that owns the resource and checks if it owns the resource. This reduces the need for traversals to find resources, and it allows resources to be claimed individually, rather than *en masse*. Furthermore, the audit can set the resource's `corrupt` flag before invoking the callback, so that it will know not to invoke the callback again if it traps. Here is a sketch of the code, ignoring encapsulation for data members:

```
Resource *ResourcePool::Allocate
            (Object &owner, ResourceCallback &c)
{
   Resource *r = dequeue a resource from the
                 free queue;
   if(r != NULL)
   {
      r->owner_ = &owner;
      r->callback_ = &c;
      r->orphaned_ = false;
      r->corrupt_ = false;
   }
   return r;
}

bool Owner::AllocateResource(void) // application code
{
   ResourcePool *p =
      SingletonObject<ResourcePool>::Instance();
   OwnerCallback *c =
      SingletonObject<OwnerCallback>::Instance();
   resource_ = p->Allocate(*this, *c);
   return(resource_ != NULL);
}

bool OwnerCallback::Claim(Resource& r) // subclass of
                                       // ResourceCallback
{
   Owner *o = (Owner*) r.owner_;
   return(o->resource_ == &r);
}

void ResourcePool::InvokeCallbacks(void) // called by
                                         // resource audit
{
   Resource *r;
   ResourceIterator i;
   for(r = FirstInUse(i); r != NULL; r = NextInUse(i))
   {
      if(!r->corrupt_)  // skip resource if callback
                        // previously trapped
      {
```

```
            r->corrupt_ = true;
            if(r->callback_->Claim(*r))
               r->orphaned_ = false;
            r->corrupt_= false;
         }
      }
   }
```

8.3.5 Distributed Resource Pools

When multiple processors share a pool of resources, the pool's audit must be distributed. The processor that owns the pool allocates the resources, handing them out to applications that run in other processors. When an application frees a resource, it sends a message to the pool. The pool's audit must modify step 3 of Section 8.3.1 to send a request to each processor that uses its resources. Each processor then responds with the resources that it is using. The requests and responses list all resources in a single message to reduce message traffic.

8.4 WATCHDOG

A WATCHDOG monitors a component by running a timer which the component must periodically reset. If the component fails, the timer expires and the watchdog resets the component [PONT01, DOUG03]. A typical purpose of a watchdog is to detect when a processor is no longer performing work. RUN-TO-COMPLETION TIMEOUT, described in Section 5.4.3, is an example of a watchdog.

Extreme systems often implement watchdogs that check the sanity of each processor. Each watchdog guards against catastrophic errors, such as a corrupt scheduler queue or a permanently masked clock interrupt, which prevent its processor from performing work. It is implemented by adjunct hardware that is coupled to the CPU. Periodically, the software must reset this watchdog. If it fails to do so, the adjunct hardware forces a reinitialization. The watchdog can be reset by the scheduler itself, or by a thread that runs in the *priority* faction.

A nonmaskable interrupt should implement this type of watchdog so that the node can capture debugging information before it reinitializes. When the watchdog causes the interrupt, it starts a timer to ensure that the node actually reinitializes. If this timer expires, the watchdog autonomously reinitializes the node. This approach guards against the node not responding to the interrupt because of a corrupt interrupt vector.

8.5 PROTECTING AGAINST GOBBLERS

A **gobbler** is an application that consumes too many resources, usually as the result of a bug. An extreme system needs to guard against gobblers so that they do not prevent other applications from performing their work.

RUN-TO-COMPLETION TIMEOUT, for example, guards against applications that enter infinite loops or that simply run locked too long. However, it does not detect all types of CPU gobblers. Consider two applications that exchange messages. A bug in these applications might result in an infinite messaging loop. In this situation, neither application runs locked too long; they merely exchange messages at such a rate that they consume an inordinate amount of CPU time.

A LEAKY BUCKET COUNTER can detect an infinite messaging loop. The counter might allow a state machine n transactions in t seconds. If the state machine enters an infinite messaging loop, its leaky bucket counter drops to zero, so it is killed.

Other types of resources, such as memory or hardware devices, are managed by pools. Earlier in this chapter, we discussed how an AUDIT returns an orphaned resource to its pool. An audit can also check how many resources an application is using. If an application is using too many resources, the audit might forcibly reclaim them. If the resource is memory, this also means killing the faulty application. If this seems too risky, the audit can generate a log instead, so that a craftsperson can decide whether to kill the application. In the case of a hardware device, however, the audit can send a message to the application, telling it to release the device. All users of hardware devices must support this message because it is also useful in other situations, such as when a device pool is empty and a higher priority application requests a device.

8.6 ESCALATING RESTARTS

The term **restart** means reinitializing part or all of a software load. Whether a restart is initiated autonomously or manually, it causes an unplanned outage. An extreme system must return to service quickly when a restart occurs. To support this, it defines different severity levels for restarts. The least severe form of restart returns a node to service as fast as possible. However, this speed entails some risk, because there are certain types of errors, such as severe memory corruptions, which it will not fix. If such an error occurs, the node will soon run into trouble again, necessitating another restart. ESCALATING RESTARTS increases the severity level of the next restart, under the assumption that the previous restart failed to correct the problem that caused the outage.

A system might define the following types of restarts, in order of most to least severe:

1. *Reboot*. The node performs a full reboot by reloading its software and reinitializing from scratch. A reboot is the only way to fix corrupt object code.
2. *Reload*. The software is not rebooted. However, all memory is released and the node is reinitialized from scratch. A reload restart fixes any corruptions in application data because the reinitialization must reload all data.
3. *Cold*. Only unprotected memory is released. WRITE-PROTECTED MEMORY is preserved so that time does not have to be spent reloading subscriber profiles, for example. A cold restart fixes any corruptions in dynamic data. However, all work in progress is lost because state machines, for example, reside in unprotected memory. Subscribers must reinitiate their disconnected sessions.
4. *Warm*. All memory is preserved, but all child threads are killed and recreated. A warm restart fixes any corruption in `Thread` subclasses while preserving work in progress. In a system where threads are lightweight and greatly outnumbered by application objects, a warm restart usually fails to fix the problem. In that case, immediately progressing to a cold restart may be preferable. However, a successful warm restart may go almost unnoticed by most subscribers. Ideally, they might only notice a degradation in response time, one that lasts for as long as it takes to kill and recreate all the child threads.

Some would-be extreme systems do not support escalating restarts. Even worse, some do not provide a SAFETY NET for traps. Any trap therefore causes an outage, and then the most stunning thing of all happens: the node takes a core dump before it deigns to return to service. For the types of large executables that are common in extreme systems, this significantly protracts the outage.

In a distributed system, a restart (or more generally, a node outage) must be immediately communicated to all other nodes so that they can take appropriate action. Each node must stop sending messages to the failed node, and each must inform its applications of the failure so that they can clean up work that involves collaboration with the failed node. Note, however, that this should not be necessary in the case of a warm restart, because it preserves all work if it succeeds.

8.7 INITIALIZATION FRAMEWORK

In most C and C++ systems, developers continuously add software to `main` in an *ad hoc* fashion, to the point where it becomes so unstructured that it would be more appropriately named `sewerMain`. The function `#includes` the world, is many pages long, contains few loops or comments, and performs work in what appears to be an arbitrary order because it does not describe initialization dependencies.

An extreme system requires an INITIALIZATION FRAMEWORK, which provides structure for `main`. To support ESCALATING RESTARTS, this framework must also support shutdown procedures that undo some, or most, of the work performed during initialization. This allows a node to be initialized or restarted in a controlled manner.

8.7.1 Registering Modules

The initialization framework defines the class `Module` to support controlled initializations, restarts, and shutdowns. Each subclass of `Module` is a singleton that represents a logical software component, often known as a **subsystem**. A module's responsibility is to initialize, restart, or shut down its subsystem.

Because modules initialize a load, they must be the first objects that `main` creates. The `Module` constructor adds each `Module` subclass to a global `ModuleRegistry`. Thus, `main` simply begins by creating each module:

```
if(SingletonObject<FirstModule>::Instance() == NULL)
   exit(-FirstModuleId);
if(SingletonObject<SecondModule>::Instance() == NULL)
   exit(-SecondModuleId);
```

And so on until `main` has created all modules. It would be preferable if `main` did not have explicitly to create all modules. However, defining the singleton instance of each module at file scope, in the hope that these singletons will be automatically initialized and registered prior to entering `main`, doesn't work. The reason is that none of the singleton instances are actually referenced anywhere, which causes C++ to optimize them out of the load. See the discussion of C++ start-up in [LAK96], however, for possible ways to circumvent this restriction.

8.7.2 Initializing a Node

After `main` has created all of the modules, it simply invokes a `Main` function defined by the singleton `ModuleRegistry`. This function does the following:

1. Initializes the load by invoking each module's `Initialize` function, whose purpose is to initialize its subsystem's global variables and create its singletons.
2. Wraps the **primordial thread**, which is the currently running thread – the one that invoked `main`. `PrimordialThread` is a singleton subclass of `Thread`. It differs from other threads in that it uses a `Thread` constructor to wrap itself. All other threads use the standard `Thread` constructor, which actually creates a native thread. The reason for wrapping the primordial thread at this point is so that it can use `Thread` functions. This allows it to begin running locked before other threads are created:

   ```
   Thread *pt =
      SingletonObject<PrimordialThread>::Instance();
   if(pt == NULL) exit(FailedToWrapPrimordialThread);
   MutexOn();
   ```

3. Invokes each module's `Start` function, whose purpose is to create any threads that a subsystem requires. Each thread subclasses from `Thread`, whose constructor creates a native thread. Under priority scheduling, if a newly created thread has a higher priority than the primordial thread, it begins to run immediately – as soon as the `Thread` constructor creates it. At this point, the newly created `Thread` object is not fully constructed, so an exception occurs if the new thread tries to invoke any `Thread` function. Furthermore, the node is still initializing, and the new thread should not interfere with this work. It must therefore go to sleep immediately, to allow time for its `Thread` object to be fully constructed. `Thread::EnterThread` (in Section 8.1.5) invokes *`sleep`* for this reason.
4. Enters the primordial thread's primary processing loop. To ensure that any exception in the primordial thread in caught by the

Thread SAFETY NET, it should be entered in the same way as all other threads, namely through the Thread::Start function:

```
pt->Start();
exit(PrimordialThreadExited); // if primordial
                              // thread returns
```

The primordial thread is the parent of all other threads. Because its death causes an outage, it must remain simple. It should not perform any application work; rather, it should limit itself to recreating killed threads. Its Enter function simply releases the run-to-completion lock and enters a simple loop which waits for SIGCHLD signals after invoking sigsuspend [GNUSIG] or which monitors child threads using the technique described in Section 13.3.5.

8.7.3 Determining Module Initialization Order

A key question is how to determine the order in which modules initialize. A simple approach is to initialize modules in the same order as they were created and registered. Under this approach, if module *B* depends on module *A*, main must create *A* before *B*. The drawback to this approach is that comments in main must describe all of the dependencies that are implicit in the code. Unfortunately, it is naive to believe that all of these comments will be added.

A more complex approach is to assign a numeric identifier to each module. When the module registers, it provides its identifier as an argument that determines its location within the Module array defined by ModuleRegistry. The primary purpose of the identifiers, however, is to allow each module to provide an explicit list of the modules on which it depends. This allows the registry to construct a directed graph that describes all module interdependencies. This dependency graph will probably be a disjoint, partial, ordering. In other words, there will be situations where there is no dependency chain between two modules. Here, any order suffices. The advantage of this approach is that it explicitly documents initialization dependencies in the code rather than in comments.

8.7.4 Restarting a Node

In addition to Initialize and Start, Module defines Restart and Shutdown functions. Note that Shutdown functions must be invoked in reverse order, and prior to, Restart functions. This is because `Shutdown` functions act as subsystem destructors, whereas `Restart` functions act as subsystem constructors.

Two forms of shutdown are useful. The first is an immediate shutdown that ends all application activity and leaves the processor in a state where it can receive maintenance commands. The second is a delayed shutdown, sometimes referred to as a **deload** operation. Here, applications continue to handle work that users have already initiated. However, no new work is accepted. In a telephone switch, for example, an immediate shutdown results in the loss of all calls. However, a delayed shutdown allows those calls to continue until they end, but the node accepts no new calls. At some point, the grace period provided by the delayed shutdown ends, and the node terminates all calls.

The purpose of `Shutdown` functions (other than when used to perform a deload operation) is to release resources (memory and threads) prior to invocation of the `Restart` function. It may be possible, however, to avoid the use of `Shutdown` functions prior to a restart. This can be done if base software can simply free the appropriate memory and kill the appropriate threads. Not only is this faster, it is safer than using operator `delete` to free application objects *en masse*. An OBJECT POOL, for example, can simply reclaim all of its blocks and free them without deleting them. True deletion, on the other hand, invokes destructors, some of whose cleanup actions are pointless before a restart. Moreover, invoking destructors is risky. If the restart occurred because a memory corruption caused too many exceptions, invoking destructors during the restart will probably lead to even more exceptions.

Therefore, to prevent pointless or risky behaviors during restarts, functions must be able to consult a global variable to determine if a restart is underway. Its value is a struct that specifies the node's state (*running*, *restart*, or *shutdown*) and the type of restart that is underway (*reload*, *cold*, *warm*, *deload*, or *none*). A parent thread, for example, might consult this variable to prevent itself from recreating child threads during the shutdown phase. Generally, any code that runs during restarts, and possibly while the node is in service, uses this variable to vary its behavior.

8.7.5 Ensuring that Initialization Succeeds

A WATCHDOG must ensure that a restart actually finishes. A thread that is created early during the restart can implement this watchdog. It simply goes to sleep. During a successful restart, the primordial thread kills the watchdog thread after initializing the last module.

Thus, if the watchdog thread actually wakes up, it causes another restart.

It may be preferable to have a payload thread, rather than the primordial thread, kill the watchdog thread. The reason for this is that a node is not truly in service until it begins to process payload work. After the primordial thread has initialized all modules, it still takes a while for all threads to start performing their work. Killing the watchdog thread too early runs the risk of not detecting a situation in which a node never starts to perform payload work.

8.7.6 Thread Observers

As a node initializes, one component may need to know when another one is in service. To support this, the initialization framework should provide a registry for thread observers. If a component needs to know when a particular thread has started to perform work, it registers as an observer of that thread by providing a callback function. When a thread starts to perform work, it informs the registry, which in turn notifies each of the thread's observers by invoking its callback function. The callback function might, for example, send a message that prompts the observer to perform work that requires the services of the observed thread.

8.7.7 Binary Database

A reload restart can take a long time. Most configuration data resides in WRITE-PROTECTED MEMORY, but a reload restart clears this memory. The node must therefore recreate its configuration data by reapplying the database transactions that originally populated its protected memory. If the database contains 100,000 subscriber profiles, recreating it takes several minutes, significantly lengthening the duration of the outage.

To speed up reload restarts, some extreme systems use a BINARY DATABASE. When a node is in service, it periodically saves a binary image of its database by dumping the contents of its write-protected memory to disk. If a reload restart then occurs, the node reinitializes its protected memory from this binary image.

This approach has one major drawback, however. While a node is dumping its binary database, it cannot allow changes to write-protected memory. Such changes could create inconsistencies in the database, because a change might alter both memory that had already been dumped and memory that had *not* yet been dumped. This would result in an image that had only committed *part* of a database transaction, hence the inconsistency. The inconsistency could be quite severe, involving stale pointers for example.

To prevent changes to write-protected memory while a database dump is in progress, `Thread::MemUnprotect` is modified to return a `bool`. During a database dump, it returns `false`. Software that modifies memory must account for this possibility and fail gracefully.

This restriction means that, during a database dump, craftspeople cannot use any administrative commands that modify the protected database. For example, they cannot create, modify, or delete subscriber profiles. Scheduling the dump during the middle of the night mitigates this annoying restriction, but it nonetheless persists for quite some time. Because the node is still performing payload work, the dump must run as a background task, so it could take several hours to complete.

Once a node has a binary image of its database, however, it can perform a reload restart far more quickly. Now another problem arises: database changes made after the dump are lost. To resolve this, the node timestamps each dump and saves subsequent database transaction commands in a file. After the node reloads the binary image, it can then bring the database up to date by reapplying these commands.

Another risk is that the database might already have been corrupt when the image was taken, in which case another restart could occur. To defend against this possibility, the next restart must revert to the previous image of the database. The timestamp associated with each image allows the node to go back in history, and the saved database commands allow it to bring the database up to date afterwards.

Where does this regression end? The answer is that the node eventually uses the original image associated with the current software release. The images are large, so the node only has disk space for a handful of them. At that point, the node deletes the oldest image to make room for the new one. However, it never deletes the original image. Thus, if a series of restarts regresses through the images taken during perhaps the last week, the previous image will be the original one.

How do we know that the original image is sane? The answer is that this image has already been tested. If a node can quickly load its data from a binary image, it can quickly install a new software release in the same way: load the new software, and then load the binary database. This binary image was created offline, by slowly applying all of the customer's database transactions to the new release. At the point, a binary image was taken and shipped to the customer for installation with the new software release.

8.8 SUMMARY

- `Thread::Start` is the entry function for all `Thread` subclasses. It provides a SAFETY NET, which catches all exceptions and signals, logs them, and invokes a virtual `Recover` function to free resources. It can then reinvoke the virtual `Enter` function that implements the application-specific entry point.
- `Thread::Start` catches exceptions by invoking the `Enter` function in a `try` block that precedes a series of `catch` blocks.
- `Thread::Start` catches signals by registering a signal handler that either throws an exception or uses `longjmp` to unwind the stack to a `setjmp` call in the `try` block. Throwing an exception is better because it destroys objects on the stack. However, many compilers and operating systems do not support throwing exceptions from signal handlers.
- To recover from stack overflows, the signal handler uses `sigaltstack` to run on its own stack.
- A LEAKY BUCKET COUNTER detects when an event occurs a specified number of times in a specified interval. Fault-handling software often uses it to trigger corrective action when a fault occurs with sufficient frequency.
- An AUDIT is a background thread which recovers orphaned resources or fixes inconsistencies in important data structures. By correcting faults, it prevents a system from eventually suffering an outage.
- Before an audit dereferences a pointer or invokes a function, it flags the underlying object as corrupt. If the operation succeeds, the audit clears the flag. If it traps, the audit knows to skip the object after SAFETY NET reenters it.
- A WATCHDOG is a sanity timer whose expiry triggers corrective action. Extreme systems use hardware watchdogs to detect hung processors.
- ESCALATING RESTARTS return a node to service quickly by performing a partial reinitialization. If the node restarts again within a short time, the severity of the reinitialization increases. A warm restart kills and recreates threads. A cold restart also clears unprotected memory. A reload restart also clears and reloads WRITE-PROTECTED MEMORY. A reboot also reloads software.

- An INITIALIZATION FRAMEWORK provides structure to `main`. Each subsystem registers a `Module` in a `ModuleRegistry` that subsequently initializes the load in an order that satisfies module dependencies. To support ESCALATING RESTARTS, `ModuleRegistry` invokes `Module::Shutdown` and `Restart` functions.
- Reinitializing WRITE-PROTECTED MEMORY with a BINARY DATABASE significantly speeds up a reload restart by avoiding the need to reapply database transactions one at a time. Creating a BINARY DATABASE offline, and then delivering it to the customer, speeds up the installation of a new software release.

9

Messaging

Extreme systems primarily consist of components that communicate with messages. An extreme system interacts with its users and administrators by receiving and sending messages. Internally, messages provide interprocessor communication and, frequently, intraprocessor communication between different applications, such as state machines. Because messages are endemic to extreme systems, their implementation must be reliable yet efficient. This chapter discusses techniques that help to achieve these characteristics. However, it does not discuss general design principles for protocols. For this topic, see [PAR00], [PAR01], and [PAR02].

9.1 RELIABLE DELIVERY

The messaging system should make every effort to deliver messages that are internal to the system, whether these messages are intra- or interprocessor. Although the messaging system cannot guarantee delivery, Reliable Delivery loses messages so infrequently that the vast majority of applications need not retransmit internal messages.

External messages, however, may require retransmission, depending on how they are transported. For example, if UDP provides transport, an application must perform retransmission. Note that applications, rather than the messaging system, typically retransmit external messages. One reason for this is that the retransmission timeout is usually protocol specific, and it may even change after each retransmission because of a back-off strategy. It is also easier for an application to cancel a retransmission timer when a response arrives than it is for the messaging system to identify the response

Robust Communications Software G. Utas
 ISBN: 0-470-85434-0 (HB)

that should cancel the timer. Putting retransmission in the messaging system often increases complexity by broadening the collaboration between applications and the messaging system.

The rest of this section focuses on internal messages, given that a protocol standard or some other specification usually defines procedures for ensuring the reliable delivery of external messages.

Many situations prevent the guaranteed delivery of internal messages. In almost all these situations, retransmitting a message is futile, which is why applications should rarely retransmit internal messages. In fact, retransmission is risky because it can easily lead to message floods during error situations.

Let's look the situations that make guaranteed delivery impossible.

1. The destination is out of service.

   When a node goes out of service, it is important to immediately inform all other nodes of this fact. When an application tries to send a message to an out-of-service node, the message `Send` function should return a value indicating that it cannot deliver the message, so that the application can react immediately.
2. The destination is unreachable because of a fault in the messaging fabric.

   Transient faults in the messaging fabric should be handled at link level, with the messaging system attempting retransmission or trying an alternative path if one is available. However, it may be impossible or impractical to inform the application that the message could not be delivered, given that applications in an extreme system send messages asynchronously.

   In most extreme systems, this situation rarely arises because the messaging fabric is duplicated, typically in a LOAD SHARING configuration (see Section 11.1.1). This provides two paths to each node, and the nodes typically duplicate the network interface cards on which they receive messages. The failure of both messaging fabrics causes a full outage, and the failure of both interface cards causes a node outage.

   Having two paths to each node in a LOAD SHARING configuration creates the possibility of message reordering, which must be avoided. A typical approach when sending IP messages, for example, is to send messages to even-numbered ports on one path and message to odd-numbered ports on the other path. This provides load balancing while preserving the order of messages sent to the same application, provided that an application does not receive messages from both even- and odd-numbered ports, which is usually a reasonable assumption. In the rare cases where this causes a problem, the sender must be able to explicitly specify the path that a message will take. This capability is also useful in other unusual situations, such as when diagnostic software wants to test a specific path.

3. The destination's message buffer is full.

   This situation should occur infrequently if message buffers are engineered for times of peak usage and if the destination implements overload controls (see Chapter 10). When a message buffer overflows, it is typically viewed as a serious error that mandates a reset of the destination, whether it is a node or thread.

   Some operating systems handle buffer overflow poorly, either by silently dropping messages or by not flagging the problem immediately. It is good practice to test how your operating system handles buffer overflows, regardless of what its documentation says about the matter.

4. The destination receives the message but encounters a software error, such as a trap or logic fault, which prevents it from processing the message.

   Retransmission is usually pointless in this situation. For example, if a state machine encounters an exception, SAFETY NET cleans it up, at which point resending the message amounts to trying to reach an out-of-service destination. However, a sender might retransmit a critical message if it knows that the receiver will be able to handle it regardless of state. Certain system control messages, such as those that inform a node that another node has gone out of service or has returned to service, fall into this category.

   When SAFETY NET cleans up an application, the recovery procedure should allow the application to send an 'I died' message to each application with which it is collaborating. These applications can then clean themselves up as well, or take other appropriate action.

5. The destination receives the message but discards it because it is applying overload controls.

   Retransmission is futile unless the overload condition has since ended. It is far more likely, however, that the overload controls are still in effect. Here, retransmission is execrable because it exacerbates the overload situation.

Guaranteed message delivery is therefore predicated on the destination being in service, reachable, and immune to software faults and overload. Because these conditions cannot be eliminated, an application must use a timer when expecting a response to a message, at least if the absence of a response could cause a fault. If the timer expires, the application takes corrective action by, for example, sending a nack to its client or releasing the resources associated with the session.

9.2 MESSAGE ATTENUATION

Carelessly designed applications can flood a system with messages, particularly during initialization or error scenarios. In an extreme system, it is unacceptable to send hundreds or even dozens of messages at once. An application that does so must be modified so that its messages are either bundled (groups of small messages combined into larger messages) or leveled (sent over a longer period).

Message bundling is a common technique when sending many messages to the same destination. Because a bundled message eliminates redundant message headers, it consumes less buffer space than a series of individual messages. It also saves CPU time because it only causes one I/O interrupt at the destination. Common bundling techniques include

1. Buffering messages for an I/O thread that periodically bundles them before sending them. This is an example of using HALF-SYNC/HALF-ASYNC for outputs rather than inputs. Its advantages are that it offers the highest degree of bundling and allows the I/O thread to send the messages gradually if there is a risk of overflowing the destination's incoming message buffers. Its drawback is that it increases latency – each message suffers a delay before it is sent.
2. Bundling messages at the end of a transaction. This approach avoids latency but provides less bundling and destination protection when different sessions send messages to the same destination.
3. Explicitly bundling messages when an application knows that it is building a group of messages for the same destination. This avoids the overhead of searching for messages that can be bundled.

Sockets use the first technique and address the latency issue by allowing the socket user to specify how long a message may be queued for output. If a delay is acceptable, the socket gathers additional outputs before it sends a message.

9.3 TLV MESSAGE

Internal messages should use a type-length-value (**TLV**) format (see Figure 9.1). A TLV MESSAGE encodes each of its parameter in three parts:

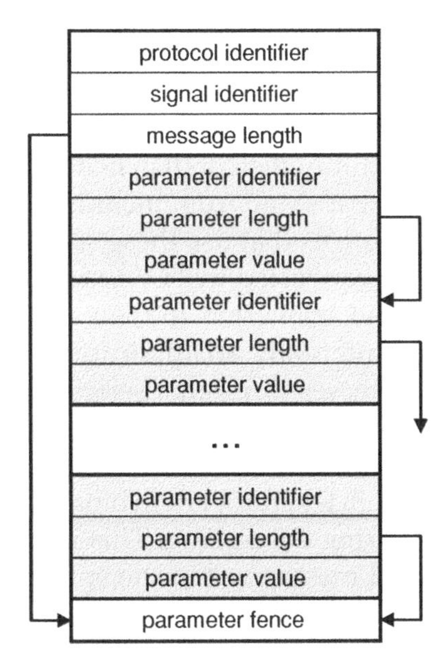

Figure 9.1 TLV message. The message header contains the protocol, signal, and aggregate length of all parameters. Each parameter specifies the length of its contents (the value). The message ends with a parameter fence (described later) that is not transmitted.

1. The parameter type, which is typically a numeric identifier.
2. The parameter's length.
3. The parameter's value, which is ideally a `struct`.

Encoding parameters in a TLV format has a number of advantages:

1. It is reasonably efficient with respect to the size of messages, particularly when parameters use binary, rather than text, encodings.
2. All parameters in a message can be traversed quickly, much faster than parameters encoded in text with <CR><LF> delimiters instead of length fields.
3. The length field supports parameters of variable length.
4. Optional information can be omitted by placing it in a separate parameter that need not be included in the message.
5. When the receiver parses the message, the length field allows it to quickly skip parameters that are not of interest.
6. Related information can be grouped into a single parameter `struct`, which promotes logical software organization and allows fast access to related information.
7. The length field allows the receiver to parse the message even if it is running a software release that does not support a parameter or one in which a parameter has a different size. In the latter case, the parameter's layout must nonetheless remain upward compatible,

meaning that all previously defined fields must still appear at the same offsets.

The TLV format is often far more space and time efficient than heavyweight formats defined by external protocol standards, including text, ASN.1, and XML. When external protocols that use these standards are required, their messages can be converted to TLV format immediately upon entering the system and converted to their native format just before exiting the system. Internally, all applications then use a TLV format. When a message passes through a series of applications that use a PIPES AND FILTERS [POSA96] or CHAIN OF RESPONSIBILITY [GHJV95] pattern, the savings can be substantial.

Some people prefer to encode parameters in text rather than binary, so that messages will be easier to read during debugging. This is a reasonable approach for infrequently used messages, but the space and time penalties often make it a poor choice for high-volume messages. In Section 9.3.3, we will discuss how parameter singletons can provide polymorphic behavior for the parameter `structs` in TLV messages. Each of these singletons can implement a `DisplayParm` function that displays a binary-encoded parameter in text format. A message trace tool (see Section 15.3.2) can then display a message in text by invoking this function on each of the message's parameters.

9.3.1 Parameter Typing

A base class can build and parse TLV messages by defining functions that add, find, and iterate over parameters:

```
typedef unsigned int   uint;
typedef unsigned short ushort;
typedef unsigned char  uchar;

typedef ushort TlvParmId;
typedef ushort TlvParmLength;
typedef int ParmIterator;

// Arbitary maximum size for a message.
const TlvParmLength TlvMaxParmLength = 1500;

struct TlvParmHeader
{
   TlvParmId     id;     // type
   TlvParmLength length; // length
};
```

```
struct TlvParmLayout
{
   TlvParmHeader header;
   char          value[TlvMaxParmLength - 1]; // value
};

class TlvMessage: public Message // partial interface
{
public:
   TlvParmLayout *AddParm(TlvParmId id,
                          TlvParmLength  length);
   TlvParmLayout *FindParm(TlvParmId id);
   TlvParmLayout *FirstParm(ParmIterator &pit);
   TlvParmLayout *NextParm(ParmIterator &pit);
};
```

Derived classes use these functions to implement those that provide PARAMETER TYPING according to each parameter's `struct`. This allows applications to read and write each parameter using a `struct`:

```
struct MyParmStruct
{
   int field1;
   int field2;
};

const TlvParmId MyParmId = 12;

class MyTlvMessage: public TlvMessage
{
public:
   MyParmStruct *MyTlvMessage::AddMyParm(void);
   MyParmStruct *MyTlvMessage::FindMyParm(void);
};

MyParmStruct *MyTlvMessage::AddMyParm(void)
{
   TlvParmLayout *pptr =
      AddParm(MyParmId, sizeof (MyParmStruct));
   if(pptr == NULL) return NULL;
   return (MyParmStruct*) &pptr->value[0];
}

MyParmStruct *MyTlvMessage::FindMyParm(void)
{
   TlvParmLayout *pptr = FindParm(MyParmId);
   if(pptr == NULL) return NULL;
   return (MyParmStruct*) &pptr->value[0];
}
```

The base class `TlvMessage` can even define function templates that provide such typing for `AddParm` and `FindParm`. Software that uses strong typing (parameter `structs`) to build and parse messages is more compact, easier to read, and more reliable than software that treats messages as byte buckets.

When a `struct` defines a parameter that appears in interprocessor messages, you must consider byte ordering when not all processors use the same endian format, and packing rules when not all processors use the same compiler. First, you need to standardize a message's endian format and perform byte flipping just before a processor of the wrong endianism sends a message. Packing is trickier. The first task is to create types such as `int8`, `uint8`, `int16`, `uint16`, `int32`, and `uint32`, and define them in terms of the appropriate `unsigned`, `char`, `short`, `int`, and `long` types on a per-compiler basis. The second task is to ensure that `TlvMessage::AddParm` aligns each parameter `struct`, with respect to its offset from the start of a message, in a way which satisfies the processor with the largest word size.

9.3.2 Parameter Fence

When an application constructs a message, `TlvMessage::AddParm` allocates space for a parameter within a buffer that the `TlvMessage` class manages. The application passes the parameter's size to `AddParm`, which therefore knows where the parameter ends.

For a parameter of variable length, a common application bug is to write beyond the end of the area allocated for the parameter. `AddParm` can detect these bugs by placing a PARAMETER FENCE, which is a pattern like `0xaaaaaaaa`, immediately after the space that it allocates for a parameter. The next time that `AddParm` is called, it can check if the previous parameter trampled any words beyond its slot. If `AddParm` detects trampling, it throws an exception to identify the culprit. Trampling after the last parameter in the message can be detected before sending the message, in the `TlvMessage::Send` function.

The use of fence patterns is common. Most heap managers, for example, use fence patterns in debug builds. It is far less expensive to detect and fix tramplers in the lab than it is to correct them after they escape to the field.

9.3.3 Parameter Template

When a `TlvMessage` is instantiated, the protocol to which it belongs should be specified. A `Protocol` singleton that adds itself to a global protocol registry formally defines each protocol. A `Protocol` singleton, in turn, has registries for singletons that define each `Signal` and `Parameter` in the protocol. A message's protocol identifier, signal identifier, and total length should appear in its header, along with its source and destination addresses.

Each `Parameter` singleton should provide a default PARAMETER TEMPLATE for its parameter's `struct`. When `TlvMessage::AddParm` is invoked, it knows the message's protocol (provided when the `TlvMessage` object was instantiated) and the parameter being added (provided as an argument to `AddParm`). This allows `AddParm` to access the appropriate `Protocol` singleton in the global protocol registry, and thence the appropriate `Parameter` singleton through the `Protocol` singleton. `AddParm` can therefore initialize the parameter with the template provided by its singleton. The following code extends that of Section 9.3.1 to show how `AddParm` supports PARAMETER FENCE and PARAMETER TEMPLATE.

```
struct TlvMsgHeader
{
   uchar  protocol;  // identifier
   uchar  signal;    // identifier
   ushort length;    // excludes header and parameter
                     // fence
};

// TlvParmFencePattern: parameter fence
// TlvParmDeathPattern: marks start of trampling
//
const uint TlvMaxMsgLength =
   sizeof(TlvParmHeader) + TlvMaxParmLength;
const ushort TlvParmFencePattern = 0xaaaa;
const ushort TlvParmDeathPattern = 0xdead;

struct TlvMsgLayout
{
   TlvMsgHeader header;  // message header
   union
   {
      // firstParm: accesses first parameter
      // bytes: accesses parameters as byte bucket
      //
      TlvParmLayout firstParm;
      char          bytes[TlvMaxMsgLength - 1];
   } TlvParmArea;
};
```

```
class TlvMessage: public Message  // partial interface
{
public:
   TlvParmLayout *AddParm(TlvParmId id,
                          TlvParmLength length);
protected:
   bool   AllocBytes(ushort count, bool &moved);
   ushort *FencePtr(void);
   void   AddFence(void);
   void   CheckFence(void);

   inline uint Align64(TlvParmLength length) const
   {
      return((((length + 7) >> 3) << 3) - length);
   };
private:
   uchar *buffer_;   // array to hold message contents
   ushort buffsize_; // size of buffer
};

TlvParmLayout *TlvMessage::AddParm(TlvParmId id,
                                   TlvParmLength length)
{
   int            size;
   TlvMsgLayout   *mptr = (TlvMsgLayout*) &buffer_[0];
   TlvParmLayout *pptr;
   bool           moved;
   int            offset;

   // Check if the fence pattern is trampled. Ensure
   // that the new parameter will fit into the buffer.
   // This includes space for its header, padding (for
   // alignment), and fence pattern.
   //
   CheckFence();
   size = sizeof(TlvParmHeader) + length +
          Align64(length);
   if(!AllocBytes(size + 2, moved)) return NULL;
   if(moved) // set by AllocBytes if buffer_ moved
      mptr = (TlvMsgLayout*) &buffer_[0];

   // The new parameter starts just after the end of
   // the message. Set its header and update the
   // message length.
   //
   offset = sizeof(TlvMsgHeader) +
            mptr->header.length;
```

```
   pptr = (TlvParmLayout*) &buffer_[offset];
   pptr->header.id = id;
   pptr->header.length = length;
   mptr->header.length = mptr->header.length + size;

   // Initialize the parameter using its template.
   // Add the fence.
   //
   ProtocolRegistry *preg =
      SingletonObject<ProtocolRegistry>::Instance();
   Protocol *prot =
      preg->Protocol(mptr->header.protocol);
   Parameter *parm = prot->Parameter(id);
   memcpy(pptr, parm->Template(), parm->Size());
   *FencePtr() = TlvParmFencePattern;
   return pptr;
}

bool TlvMessage::AllocBytes(ushort count, bool &moved)
{
   TlvMsgLayout *mptr = (TlvMsgLayout*) &buffer_[0];
   uchar        *buff;
   ushort       oldlen;
   ushort       newlen;

   //  If the buffer can't hold COUNT more bytes,
   //  extend its size.
   //
   moved = false;
   oldlen = mptr->header.length;
   newlen = sizeof(TlvMsgHeader) + oldlen + count;
   if(newlen > buffsize_)
   {
      buff = (uchar*) realloc(buffer_, newlen);
      if(buff == NULL) return false;
      buffer_ = buff;
      buffsize_ = newlen;
      moved = true;
   }
   return true;
}

ushort *TlvMessage::FencePtr(void)
{
   // Return a pointer to the parameter fence.
   //
   TlvMsgLayout *msg = (TlvMsgLayout*) buffer_[0],
```

```
   int offset = sizeof(TlvMsgHeader) +
                msg->header.length;
   return (ushort*) &msg->TlvParmArea.bytes[offset];
}

void TlvMessage::AddFence(void)
{
   // Add the fence to the end of the message.
   //
   bool moved;
   if (!AllocBytes(2, moved)) return;
   *FencePtr() = TlvParmFencePattern;
}

void TlvMessage::CheckFence(void)
{
   // If the fence has been trampled, throw an
   // exception after marking the location where
   // trampling began.
   //
   if(*FencePtr() != TlvParmFencePattern)
   {
      *FencePtr() = TlvParmDeathPattern;
      throw MessageTrampledException();
   }
}
```

A parameter template acts as a constructor for its parameter's `struct` or, to put it another way, a `Parameter` singleton acts as a CONCRETE FACTORY [GHJV95] for parameters that actually appear in messages. Because a message can be interprocessor, it cannot contain parameter objects. However, the combination of `Parameter` singletons and `structs` provides a close approximation to placing true objects in messages. Using the parameter identifier (the `id` field of `TlvParmLayout`) to select the singleton provides polymorphic behavior.

Initializing a parameter with a default template takes extra time, but a block copy operation (`memcpy`) does this efficiently, and the benefits outweigh the cost. First, a block copy operation is actually more efficient than initializing fields to default values individually. Second, the template relieves applications of actually having to set default values, which makes their software simpler. Finally, the template ensures that all fields actually contain legal values: there is no risk, for example, that a field defined by an `enum` type will be out of range. A field may be initialized twice, but at least all of them *will* be initialized.

9.3.4 Parameter Dictionary

Some protocols are rich in the sense that their messages can contain many parameters. When applications process such messages, they invoke `FindParm` many times, to look for all parameters that are of interest. This results in many calls to `FirstParm` and `NextParm`, which `FindParm` uses as iterators. It may therefore be desirable to speed up `FindParm`.

A PARAMETER DICTIONARY significantly speeds up `FindParm` by providing a fast lookup table for each parameter. Parameter identifiers index the table. When a message arrives, the table is built by iterating through the parameters using `FirstParm` and `NextParm`. A subclass of `TlvMessage` that does this can then override `FindParm`, quickly returning a pointer to a desired parameter by consulting the table. This allows applications to process parameters efficiently, in whatever order makes sense to application logic rather than in the order that the parameters actually appear in the message.

9.4 ELIMINATING COPYING

The frequent copying of messages can significantly degrade capacity in message driven systems. This section discusses techniques that reduce the frequency of message copying.

9.4.1 In-Place Encapsulation

In many protocols, messages undergo encapsulation when passed to a lower layer in the protocol stack. The amount of data added during encapsulation is usually small when compared to the application payload that is being wrapped. Thus, if encapsulation requires copying the payload into a new buffer, a lot of time is wasted.

To support efficient encapsulation, the payload portion of the message should be constructed at some offset from the beginning of the buffer. The size of the offset is determined by how much space will be required to prepend one or more headers during encapsulation, as shown in Figure 9.2.

Finally, applications should be allowed to specify the size of the original buffer. This often eliminates the need to copy a buffer's contents to a larger buffer when an application repeatedly invokes [illegible]

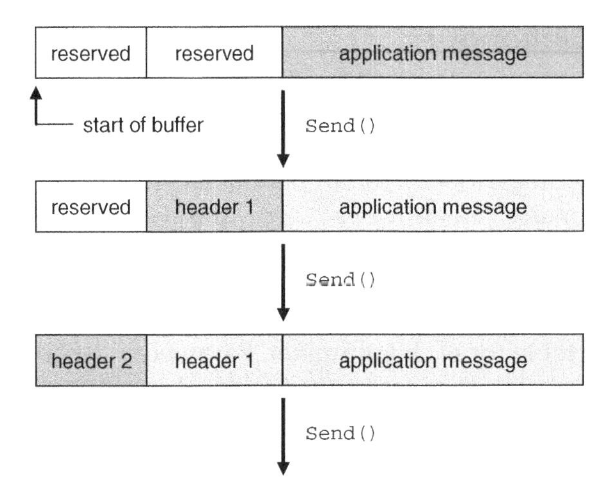

Figure 9.2 In-place encapsulation. The application layer builds its message at an offset from the start of the buffer. This provides space for lower layers to prepend headers as the message travels down the protocol stack, thereby avoiding the cost of copying the message within each layer.

9.4.2 Stack Short-Circuiting

The function `TlvMessage::Send` improves performance if it bypasses the operating system or IP protocol stack when it sends a message. To do this, it must analyze the message's source and destination addresses to determine which of the following scenarios applies (see Figure 9.3):

1. *Intra-thread*: a message to an application that runs under the same thread.

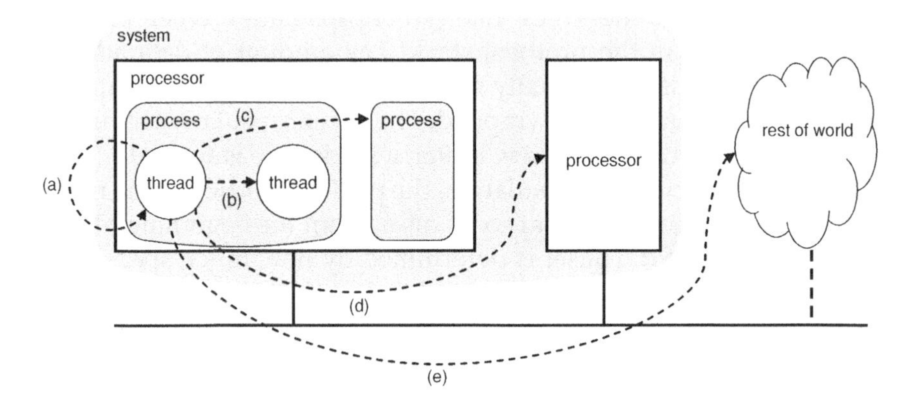

Figure 9.3 Messaging scenarios: (a) intra-thread, (b) intra-process, (c) intra-processor, (d) intra-system, (e) inter-system.

2. *Intra-process*: a message to an application that runs under a different thread within the same process.
3. *Intra-processor*: a message to an application that runs under a different process within the same processor.
4. *Intra-system*: a message to an application that runs in a different processor within the same system (that is, where the destination processor is managed by the same maintenance software as the source processor).
5. *Inter-system*: a message to an application that is external to the system.

Standards usually specify how to send inter-system messages, in which case there is no possibility for optimization. We will therefore look at the other cases.

An intra-system message should usually travel over the IP stack. There are two reasons for this. First, the system might eventually have to be decoupled as the result of evolution in standards. If this occurs, an intra-system message could become inter-system. IP is therefore a good choice because it is a common standard for inter-system communication. Second, any alternative to IP is likely to be proprietary. The system's software will be more portable if it simply uses IP instead of a custom messaging layer provided by a distributed operating system.

This being said, you need to stress test your IP stack, regardless of whether you implement it yourself or use one that your operating system provides. Two things to test are buffer overflows and timeouts. For the purposes of an extreme system, some IP stacks take too long to report overflows or to detect that a destination is out of service. If you cannot modify the stack to fix such problems, you will need to build a wrapper for it.

When using IP for intra-system messaging, the question is whether to use UDP or TCP. UDP is far more efficient, which makes it the preferred choice. On the other hand, TCP is more reliable because it provides retransmission. Thus, when reliability is critical, TCP might be a better choice. However, as discussed in Section 9.1, it is usually undesirable to retransmit internal messages. Consequently, internal messages should rarely use TCP for transport. Furthermore, the length of time that TCP takes to decide that a destination is unreachable, and to return a socket error, may be too long for extreme systems. Therefore, if UDP is front-ended by logic that knows when a destination node is out of service, it should suffice for the majority of applications.

Intra-thread and intra-process messages offer the greatest opportunity for reducing the cost of messaging. In these scenarios, a message can be placed directly on an internal work queue that is serviced by the destination's invoker thread. If this invoker thread has gone to sleep 'forever' because its work queue was empty, it must be signaled so that it will wake up and handle the new work. Frequently, however, the destination invoker thread will only be sleeping for 'zero time'. That is, it will not be sleeping because its work queue was empty, but because it voluntarily scheduled itself out when it was at risk of exceeding its RUN-TO-COMPLETION TIMEOUT timeslice. In this case, it need not be signaled: it will simply handle the new work the next time it runs.

Some systems further optimize intra-thread and intra-process messaging by directly invoking the destination's `Handle` function so that it will process the message immediately. This technique, however, has a number of drawbacks:

1. It bypasses the message ordering and prioritization provided by the destination's work queue.
2. If the destination traps while processing the message, the trap occurs in the wrong processing context. This makes it difficult for SAFETY NET to clean up the correct objects.
3. It combines separate transactions into one extended transaction. This introduces a risk of the source's invoker thread being killed for exceeding its RUN-TO-COMPLETION TIMEOUT.
4. It causes deeper nesting of function calls, which can lead to a stack overflow.

Therefore, directly invoking the destination's `Handle` function is rarely advisable. It may be reasonable when the destination will only perform a small amount of work, but the messaging system is unlikely to know when this will be the case, which means that the sender will need to tell it when to use this technique.

The final scenario is the intra-processor (but inter-process) message. Here we are dealing with different USER SPACES, so the simplest approach is again to use the IP stack. As discussed in Section 7.2.8, the primary justification for user spaces is to run different software loads on the same processor in a small system configuration. Consequently, efficiency is not a primary concern, so the overhead of the IP stack is unimportant. The IP stack will itself realize that the message is intra-processor, and will therefore perform its own short-circuiting.

9.4.3 Message Cascading

Sometimes an application needs to send the same message, perhaps with slight modifications, to many destinations (see Figure 9.4). In

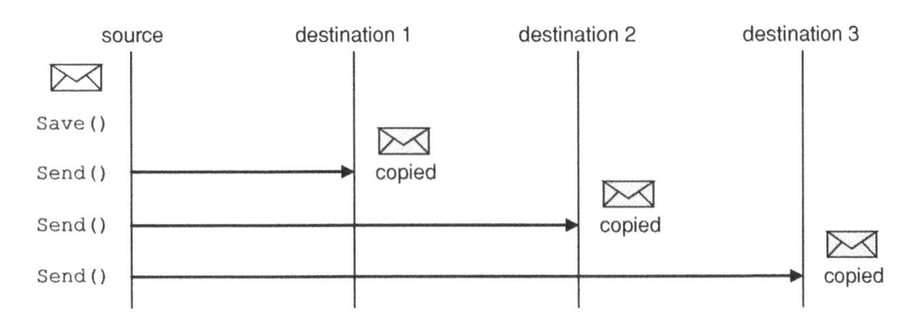

Figure 9.4 Message cascading. An application saves a message before sending it to multiple destinations. This prevents stack short-circuiting from moving the message to an interprocessor destination. The message must be copied when it is sent, but at least the sender does not have repeatedly to reconstruct it.

such cases, the application should be able to prevent STACK SHORT-CIRCUITING from moving the message directly to an intraprocessor destination. If the message moved, the application would have to reconstruct it repeatedly, because it is too complicated and risky for multiple destinations to share a message. A shared message requires reference counting. It must enforce a read-only policy, which is difficult to do. And a stub message, referencing the shared message, must be queued at each destination. This introduces even more complexity and processing costs when applications access the message through a stub rather than directly.

It is therefore desirable to provide a `Save` function for messages. When a message is sent, it can normally be moved to its destination (if intraprocessor) or destroyed immediately after queuing its buffer at I/O level (if interprocessor). However, if a message has been saved, the message's `Send` function must copy the buffer instead of appropriating it.

A simple counter can implement the `Save` function: if the counter is nonzero, the message has been saved, so it cannot move to the destination. When the application no longer requires the message, it invokes an `Unsave` function that destroys the message when the counter drops to zero.

`Save` and `Unsave` are also useful when an application may later need to retransmit a message. Although a buffer must be copied when sending a saved message, the block copy operation used for this purpose will be more efficient than forcing the application to reconstruct the message.

9.4.4 Message Relaying

When an application is part of a PIPES AND FILTERS or CHAIN OF RESPONSIBILITY pattern, it sometimes receives a message, decides

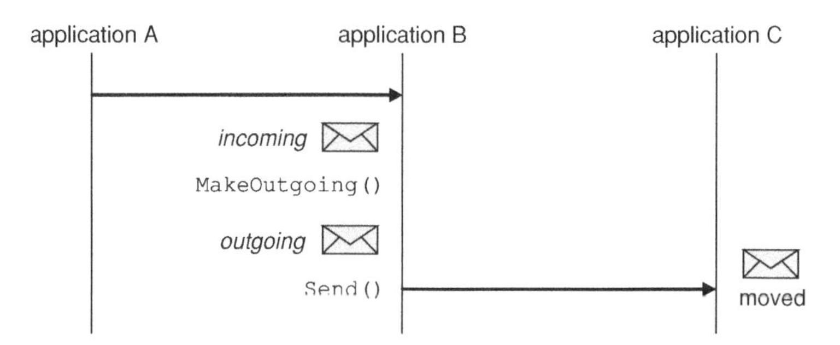

Figure 9.5 Message relaying. To forward a message from application A to application C without copying it, application B must change the message's role from incoming to outgoing because an incoming message (a) does not support a `Send` function and (b) is read-only, whereas its source and destination addresses must change when it is sent. Application B has not saved the message, so `Send` moves it to application C.

that it has no work to do, and sends the message to the next application in the chain (see Figure 9.5). In this situation, the ability to relay the message to its next destination, without copying it, saves time. Only the message's source and destination addresses change.

Relaying a message without copying its buffer is simple. As long as the message has not been saved, its buffer can be appropriated when sending the message.

However, what about the `TlvMessage` object that wraps the buffer? Can it also be reused? Here there are two situations to consider. First, the application must be able to take an incoming `TlvMessage` and invoke the `Send` function to relay it. Second, in the intraprocessor case, it is desirable to move the entire `TlvMessage` object to the destination, rather than only its buffer. This way, the destination does not have to construct another `TlvMessage` object when it receives the buffer.

Reusing a `TlvMessage` object in this way can be challenging, however. The reason is that incoming and outgoing messages may have different subclasses, even if their buffers' contents are identical. Different subclasses might be used because incoming messages should be treated as read-only. It is undesirable for an incoming message to support functions like `AddParm` and `Send`. The availability of such functions for incoming messages opens a system to inadvertent errors or deliberate kludges.

Of course, there is an alternative to using subclassing to enforce restrictions on incoming messages. If `TlvMessage` defines a `direction` attribute (`incoming` or `outgoing`), functions like `AddParm` and `Send` can check this attribute. Many people would find this acceptable, but those with strong object-oriented sensitivities might find it objectionable because it uses an attribute to implement behaviors that could be separated through subclassing.

Whether subclassing or an attribute implements restrictions on incoming messages, `TlvMessage` needs to define the functions `MakeOutgoing` and `MakeIncoming`. These functions either change the `direction` attribute or morph the message to a suitable subclass, as discussed in Section 4.4.4. Protocol-specific message subclasses may need to override these functions in order to fix up data that would otherwise be missing after a message changes its direction. At the cost of some additional complexity, these functions save time by allowing the original `TlvMessage` object to pass from one application to the next.

9.4.5 Eliminating I/O Stages

Reducing the number of stages through which an incoming message must pass can significantly improve a messaging system's performance. An incoming message may have to traverse as many as four stages before it is processed! Here, we are referring to interprocessor messages, for which STACK SHORT-CIRCUITING is impossible. Indeed, the purpose of short-circuiting is to bypass one or more stages. For an incoming IP packet, the stages are

1. An **ISR** (interrupt service routine).
2. The **network thread**, which sometimes goes by the name `tnettask`.
3. An I/O thread that receives messages and places them on a work queue.
4. An invoker thread that services the work queue.

Depending on the type of input, you may need to modify the operating system to eliminate some of these stages. This is indeed the case in our IP example, but it is not always so. It depends on whether the operating system provides a protocol stack that handles the inputs.

Let's look at some ways to process an incoming IP packet (see Figure 9.6). In each of them, an incoming message causes an interrupt, after which an ISR is invoked to receive the message. The ISR is, therefore, the first stage in each example.

1. *Four-stage input*. The ISR queues the packet for the network thread, which is the second stage. If a message has been segmented into multiple packets, the network thread assembles them. It then

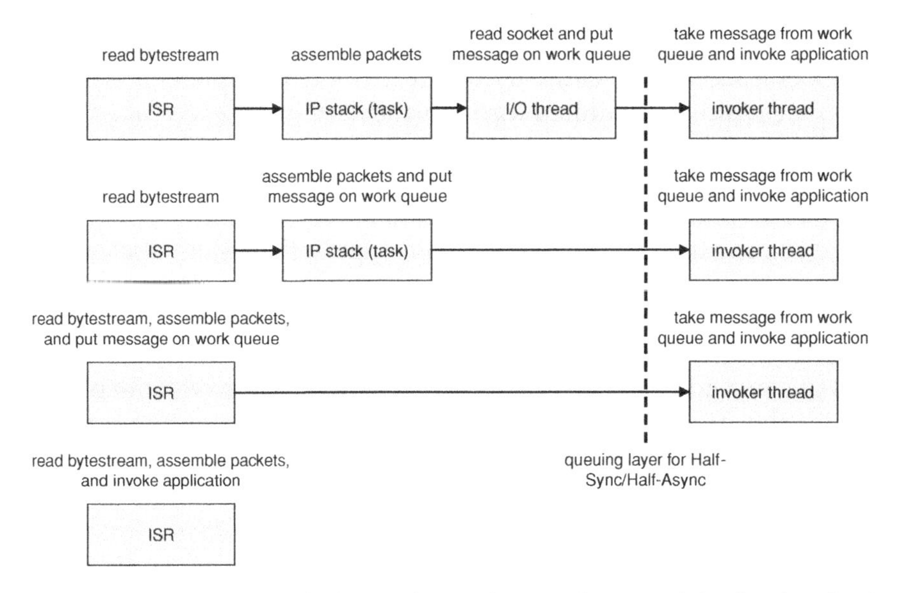

Figure 9.6 Eliminating I/O stages. The default pattern contains four levels of scheduling. This can be reduced to as little as one stage if it is possible to modify parts of the operating system.

delivers them to an I/O thread that uses a blocking call like `recvfrom` to wait for messages. The I/O thread is the third stage. Finally, the I/O thread places the message on a work queue that is associated with an invoker thread. The invoker thread is the fourth stage, and it periodically wakes up to service its work queue.

If an I/O thread receives messages in different protocols, it must delegate some of its work to an **input handler**. Each input handler supports a different protocol. It takes an incoming message, wraps it with a suitable object, and determines the work queue on which to place it. How to wrap a message, and where to queue it, are based on the message's protocol. For messages arriving on the IP stack, the port typically determines the protocol, and so a different input handler registers against each IP port on which messages can arrive. These input handlers reside in an **IP port registry**.

2. *Three-stage input*. This approach merges the functions of the network thread and I/O thread into the network thread. It eliminates the I/O thread because it is inefficient to pull messages from the network thread. Instead, it modifies the network thread to push messages directly into the application input handlers that, in the four-stage approach, are invoked by the I/O thread. The network thread therefore requires access to the IP port registry (of input

handlers). It can then queue a message directly on an invoker thread's work queue by pushing it into the appropriate input handler, thereby avoiding the overhead of an I/O thread. This technique must also modify the network thread to use SAFETY NET in case an input handler (now running out of the network thread) traps.

3. *Two-stage input*. This approach bypasses the network thread to eliminate its overhead. When a UDP packet is destined for an application, the network thread adds little value because packet reassembly is not required. Therefore, if the ISR has access to the IP port registry, it can queue a message directly on an invoker thread's work queue by pushing it into the appropriate input handler. This technique involves modifying the ISR. Furthermore, the input handlers must be bulletproof, because a trap in an ISR causes a crash in some systems. Finally, both the ISR and invoker thread manipulate work queues. The invoker thread must therefore provide critical region protection by disabling I/O interrupts while it is accessing its work queue.
4. *One-stage input*. This approach eliminates the invoker thread: all of the work associated with a message runs at interrupt level, in the ISR. This is the most efficient technique of all, but it is rarely appropriate for the types of extreme systems that are the topic of this book. The reason is that, by eliminating HALF-SYNC/HALF-ASYNC, it makes it impossible to prioritize work, a capability that is needed to support the overload controls described in Chapter 10. However, one-stage input may be appropriate for an application that is essentially firmware, such as simple IP routing or ATM switching. A system that contains such an application would use a multistage approach to implement its control plane but might implement its data plane entirely at interrupt level.
5. *Zero-stage input*. Given that one-stage input is primarily suited to applications that are firmware, the final optimization is to do everything in hardware. The data planes of some routers and switches do indeed take this approach, but such applications are outside the scope of this book.

In an extreme system that runs on an off-the-shelf operating system, four-stage input is the standard way to receive IP packets in a HALF-SYNC/HALF-ASYNC configuration, but if you need more throughput, and have the option of modifying the network thread or I/O ISR, you should consider one of the more lightweight techniques. For each I/O stage that you remove, the costs of context switching and message copying, from one stage to the next, vanish.

9.5 ELIMINATING MESSAGES

Each message takes time to build, send, receive, and parse. Even if we eliminate message copying, these costs remain. Therefore, eliminating entire messages is even better than eliminating copying.

External messages must obey protocol standards, so eliminating them is usually impossible. Protocols that are internal to the system, however, should be designed to avoid unnecessary messaging.

9.5.1 Prefer Push to Pull

When an application pulls data using a request-response message sequence, consider using an unsolicited message to push the data instead. Because this eliminates a message, it often improves capacity. This technique often involves using HALF-OBJECT PLUS PROTOCOL, which was discussed in Section 6.3.5. The master object pushes data to an object that provides a copy, allowing applications to access the data with a function call (see Figure 9.7).

When data is pushed rather than pulled, the question is whether or not this will reduce the total number of messages. In a request-response sequence, the total number of messages is twice the number of requests. With indication messages, the total number of messages corresponds to the number of times that the data changes. Thus, if for every n data changes, there are fewer than $n/2$ requests, the request-response sequence actually results in fewer total messages.

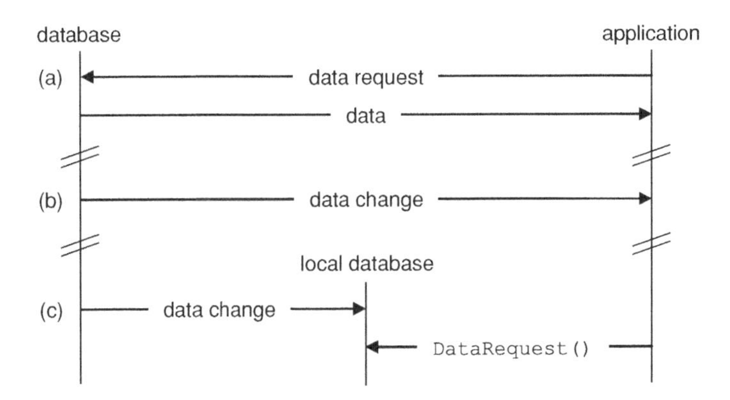

Figure 9.7 Pull versus push. In (a), an application pulls data from a database. In (b), the database notifies the application of data changes. In (c), the database uses Half-Object Plus Protocol to push data changes to a local copy of the database so that the application can use function calls to access data.

To take a simple example, say that a system monitors the status of a component by periodically sending it an 'Are you alive?' message and waiting for a 'Yes, I'm alive' response. Replacing these with an unsolicited 'I'm alive' message improves efficiency. Even better would be an 'I died' message, assuming that there is a way to generate one when a software component fails. If a thread dies and is not recreated, then its parent could generate this message while handling the `SIGCHLD` signal.

Pulling data has one advantage over pushing it. When a processor is busy, it may want to limit how much time it spends updating data used by low-priority applications. If it pulls the data, it controls the amount of time spent on updates. However, if updates are pushed, the processor must handle them unless some form of catch-up scheme exists. Handling the updates subverts the processor's ability to prioritize work, and the catch-up scheme adds complexity. In such a situation, pulling the data is usually a better choice.

9.5.2 No Empty Acks

As discussed in Section 9.1, retransmitting internal messages is usually undesirable. This means that you can remove empty acknowledgments. Simply assume that work delegated elsewhere will succeed, and only send a nack if it fails. The application that requested the work can then handle the nack asynchronously, as shown in Figure 9.8. In many cases, the application needs to deal with delayed failures anyway. The processing of an immediate nack or a delayed failure is usually identical, in which case the two need not even be distinguished. A protocol's initial design often contains empty acks whose elimination would improve the system's capacity.

As an example, consider a request to set up a connection between two ports on a switch. This request might fail immediately, perhaps because of blocking or some other problem within the transport fabric. In the vast majority of cases, however, it will succeed. Furthermore, even if a connection is successfully established, it might later fail as the result of a hardware or software fault. Capacity can therefore be improved by eliminating acks to connection setup requests. The requester only receives a message if the connection cannot be set up or if it later fails. In both cases, the requester receives the bad news asynchronously, and deals with it in the same way.

Eliminating empty acks has one drawback. If a request is lost, the requester will assume that it succeeded. Even with RELIABLE DELIVERY, there are situations in which delivery will fail (see Section 9.1). These situations must be assessed before eliminating acks.

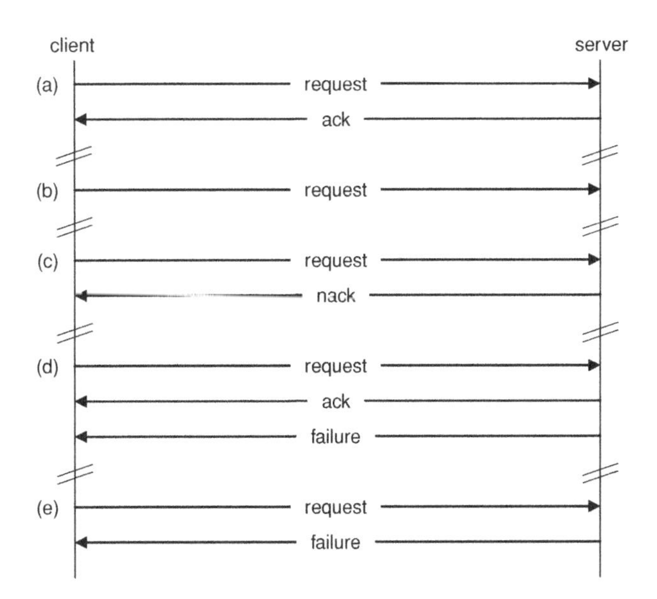

Figure 9.8 Eliminating acks. In (a), the server sends the client an empty ack. The server eliminates the ack in (b), and the client assumes that the request succeeded. In (c), the request immediately fails, so the server sends a nack. When the request creates a session, a failure can occur later, turning (a) into (d). If the server eliminates the ack, a failure turns (b) into (e). Because (c) and (e) are similar under asynchronous messaging, the client can handle both of these failure scenarios in the same way.

9.5.3 Polygon Protocol

When a request spawns a chain of requests, a POLYGON PROTOCOL eliminates intermediate acks. In such a protocol, the last component in the chain sends an ack to the component that made the original request. If a request fails in the middle of the chain, each component sends a nack to the previous one to allow cleanup of the partially satisfied request.

Figure 9.9 illustrates the use of a polygon protocol to set up an ATM connection. The application request goes to the ATM fabric node, which forwards it to the egress access node, which forwards it to the ingress access node, which finally responds to the application.

9.5.4 Callback

When the recipient of a message does not need to perform much work, replacing the message with a CALLBACK significantly increases capacity if the message occurs frequently.

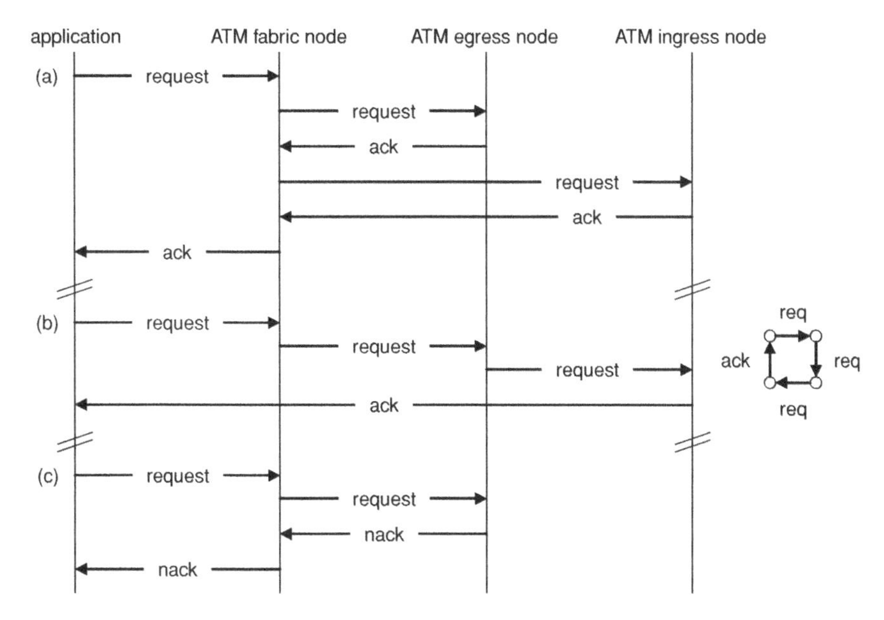

Figure 9.9 Polygon protocol. In the original protocol of (a), the fabric node coordinates the setup of an ATM connection. In (b), a polygon protocol eliminates intermediate acks. In the failure scenario of (c), intermediate nacks are required to clean up previous work.

A callback involves two objects: a subscriber object, *A*, and a publisher object, *B*. When *A* wants to be informed of an event that occurs in *B*, *A* registers with *B*. When the event occurs, *B* invokes a function (the callback) on *A*. When different types of subscriber need to register with a publisher, a BRIDGE [GHJV95] implements the callback to decouple the publisher from its heterogeneous subscribers.

Many systems use callbacks extensively. The subscriber and publisher roles mean that callbacks are a common way to implement OBSERVER [GHJV95]. Many timer services use callbacks to inform applications of timeouts. In some cases, asynchronous messages implement subscription requests, but callbacks implement the responses to those requests (the publications).

Unfortunately, callbacks suffer from a number of drawbacks:

1. If a callback traps, the trap occurs in the publisher's context rather than in the subscriber's context. In Section 8.3.1, we discussed how a resource pool AUDIT uses callbacks to prompt applications to claim in-use resources, but in Section 8.3.4, we discussed the need to deal with traps in these callbacks.
2. If the subscriber runs in an application framework, the callback bypasses this framework because the work runs in the publisher's context. This is similar to a problem in Section 5.5.3, which pointed

out that a drawback of synchronous RPCs is that responses arrive directly within application code rather than top down, through a session processing framework.

3. If the subscriber and publisher objects run in different scheduler factions (or at different priorities), the work associated with the callback runs in the wrong faction.

Because of these drawbacks, the following criteria restrict the use of callbacks:

- The publisher must use SAFETY NET to recover from a trap in a callback.
- The work performed by a callback should be simple. This reduces the risk of the callback trapping, and it avoids performing too much work in the wrong faction.
- The work performed by a callback must not require capabilities from the subscriber's application framework.

If these criteria cannot be satisfied, a message must replace the callback. The publisher can send the message, or, if different subscribers require different messages, the callback can send the message instead of performing the actual work.

9.5.5 Shared Memory

Placing data in shared memory eliminates a significant number of intraprocessor messages. The overhead of using messages to access data is significant, which is why extreme systems often prefer WRITE-PROTECTED MEMORY to USER SPACES.

Data that is not protected also benefits from being shared. Sharing is automatic among applications that run in the same process, which is why extreme systems minimize their use of processes or even run all threads under a single process. Under multiple processes, however, shared objects must reside in a shared memory segment. A `SharedObject` class can provide this capability, just as `ProtectedObject` (see Section 7.2.7) provides it for protected memory. It overrides operators `new` and `delete` to allocate and free memory in a global shared segment. This segment comprises a separate heap, although a shared object is typically allocated from an OBJECT POOL, which is created from this heap during system initialization.

9.6 SUMMARY

- Although the messaging system must provide RELIABLE DELIVERY, a number of situations preclude guaranteed delivery or a guaranteed response. In these situations, however, an application should usually not retransmit an internal message when its response timer expires.
- Applications must send messages gradually or bundle them to avoid flooding the system with messages.
- Use a TLV format for internal messages. It improves capacity and supports PARAMETER TYPING, which defines each parameter as a `struct`.
- Use PARAMETER TEMPLATE to ensure that each parameter field is initialized.
- Use PARAMETER FENCE to detect trampling during message construction.
- To improve capacity, eliminate message copying. Even better, eliminate entire messages.

10

Overload Controls

When an extreme system receives more work than it can handle, it must not thrash or crash. Its throughput must increase until it reaches and stays close to some maximum. **Overload controls** guarantee this behavior, which is illustrated in Figure 10.1.

Each processor in an extreme system must implement overload controls, which are either reactive or proactive:

- A reactive overload control triggers when a processor receives more work than it can handle. It prioritizes work to postpone or discard some of it.
- A proactive overload control prevents too much work from arriving in the first place. It does so by throttling other processors, preventing them from sending more work to an overloaded processor.

Reactive overload controls are always required. Even when it is possible to throttle other processors proactively, it is risky to trust them to perform such throttling.

Proactive overload controls eliminate the overhead of having to receive and discard work in the first place. A processor that implements them will therefore improve its capacity.

For a survey of overload control patterns, see [HAN00]. Many of the techniques that we are about to discuss appear in [MESZ96], which maps to our treatment as follows:

- The need to SHED LOAD when nearing overload has already been mentioned.
- FINISH WORK IN PROGRESS is covered under FINISH WHAT YOU START (Section 10.1).

Robust Communications Software G. Utas
 ISBN: 0-470-85434-0 (HB)

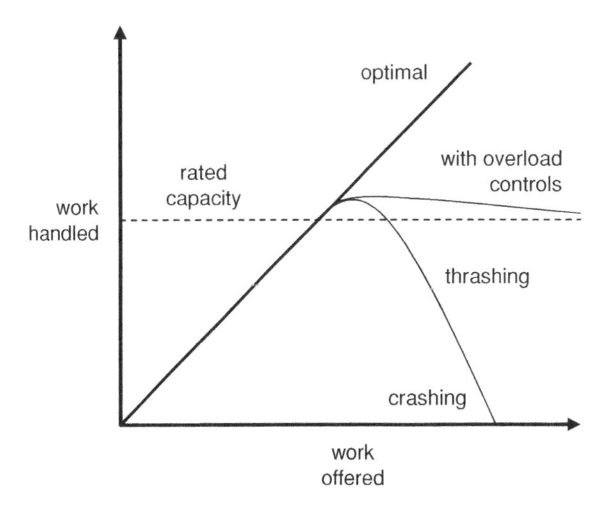

Figure 10.1 Throughput and overload. When presented with more work than it can handle, a system that lacks overload controls thrashes and eventually crashes. With overload controls, throughput tapers off gradually.

- FRESH WORK BEFORE STALE and MATCH PROGRESS WORK WITH NEW are covered under DISCARD NEW WORK (Section 10.2).
- WORK SHED AT PERIPHERY and LEAKY BUCKET OF CREDITS are covered under THROTTLE NEW WORK (Section 10.4).

The techniques in this chapter are *intrasystem* in nature. They allow a system to protect itself from the external world unilaterally, and its processors to protect themselves from each other. Many networks also adopt standards for *intersystem* overload controls, which are often known as **congestion controls**. We do not cover them here because they are network specific. For example, see [HAN99b] for a discussion of congestion controls used in telephone networks.

10.1 FINISH WHAT YOU START

Assume that a processor can handle n jobs but that it receives $n + m$ jobs. In a nonextreme system, the processor does $n + m$ jobs poorly. Thrashing and latency start to increase noticeably after n jobs, a common symptom being message timeouts. Throughput degrades as more work arrives. Eventually a crash is likely, perhaps because message buffers run out.

When faced with this situation, an extreme system must do n jobs and reject m jobs. Throughput must remain constant at n jobs. The question is that of which jobs to reject.

The foremost principle in overload controls is to handle work in progress before new work. Each message that enters a processor must be classified as new work (the creation of a new session) or progress work (a message to an existing session). Examples of new work include adding a subscriber, registering a subscriber, and setting up a session or connection. Examples of progress work include deleting a subscriber, handling a message to a session that is already underway, and forwarding packets over an existing connection.

There are a number of ways to distinguish new work from progress work:

1. Look at the message's signal. In SIP, for example, a REGISTER or INVITE represents new work, whereas a 200 OK, CANCEL, or BYE represents progress work.
2. Determine whether a message is destined for an existing processing context, such as a state machine that is not in its initial state. Such a message represents progress work, whereas a message that creates a new processing context represents new work. The response to a subscriber profile download request, for example, is a progress message because it arrives at a context that is in some kind of pending state.
3. Add a field to message headers to specify whether a message represents new work or progress work. This requires the cooperation of other nodes and their applications so that messages arrive with this field properly set. However, it is the often the best solution because it allows an application to specify that a critical message be treated as progress work when it might otherwise be classified as new work. An IP router, for example, might classify all control packets as progress work.

After a message has been classified, it is placed on an appropriate work queue. At a minimum, each processor needs two work queues, one for progress work, and one for ingress (new) work. Messages on the ingress queue are not serviced until the progress queue is *empty*. When the processor is running at peak capacity, this ensures excellent service (minimal latency) to sessions that are already in progress. Sessions receiving progress work therefore finish earlier, which frees up some CPU time for new work. The overriding idea is that it is dangerous to accept new work if there is any chance that progress work will not be handled.

In addition to the progress and ingress work queues, two other queues should be considered (see Figure 10.2). The *immediate* queue is the highest priority queue. Its purpose is to support RUN-TO-COMPLETION CLUSTERS, as defined in Section [illegible]. The egress queue

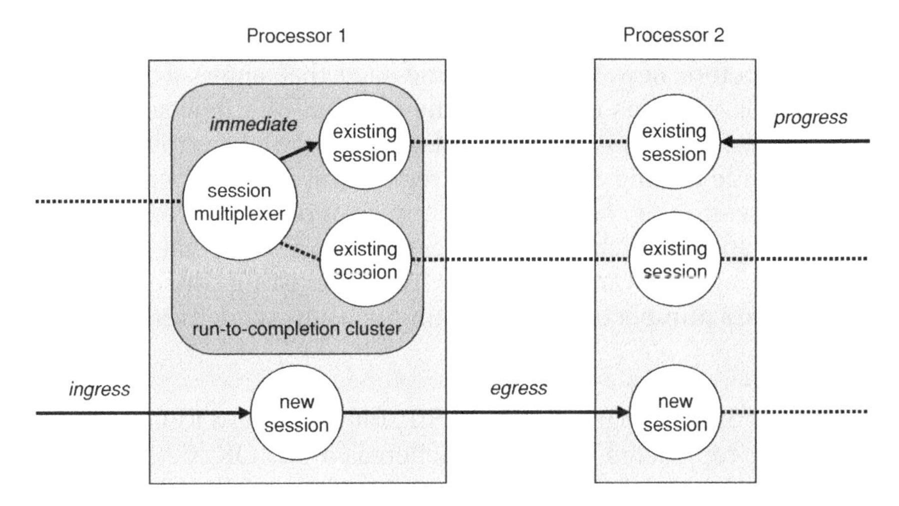

Figure 10.2 Message priorities. An ingress message creates the incoming half of a session, and an egress message creates the outgoing half of a session. A progress message is one to an existing session. An immediate message is one within a run-to-completion cluster.

has lower priority than the progress queue but higher priority than the ingress queue. It contains initial messages that session initiators send to session recipients. These messages are new work for the processor that receives them but progress work for the system as a whole. Given that the system has already invested time in creating the incoming half of the session, it should give more priority to creating the outgoing half of the session than it does to creating new sessions. When a processor incorporates these additional queues, the same principle holds: do not service work in a lower priority queue until all higher priority queues are empty, with the overall priority scheme being *immediate*, *progress*, *egress*, and *ingress*, in descending order.

10.2 DISCARD NEW WORK

When the ingress queue gets too long, stale new work is discarded, and new work may be discarded as soon as it arrives. If possible, work should be classified and discarded at I/O level to minimize the time spent on it.

When the number of message buffers on the free queue drops below a threshold, new work is immediately discarded upon arrival. Because progress work is handled before new work, the progress queue is much shorter than the ingress queue when a processor is running at peak capacity. During such times, the progress queue might contain, say, 20 messages, but the ingress queue might contain as many as 1000. An adequate number of message buffers must be reserved for progress work, as it is unacceptable to lose progress work because new work requests have used all the buffers. Therefore, when the number of buffers on the free queue falls below a threshold, new work is immediately discarded.

An estimate for the minimum length of an ingress queue is determined by

- how long a message can be queued before it times out;
- the percentage of CPU time allotted to the faction that handles the work, and
- the cost of handling each message *plus* any follow-up messages that occur within the timeout period.

The formula is

$$\frac{(\text{timeout} \times \text{faction percentage})}{(\text{cost of new work} + \text{cost of follow-up work})}$$

This allows the system to handle all the new work if it receives no progress work during the timeout period.

For example, say that

- the average user will hang up after waiting for dial tone for 5 seconds;
- call processing software receives 85% of the CPU time;
- accepting a call and providing dial tone takes 2 milliseconds, and
- receiving the digits and setting up the call takes 8 milliseconds.

Based on these criteria, the minimal length of the ingress queue would be

$$(5000\,\text{msec} \times 0.85)/(2\,\text{msec/msg} + 8\,\text{msec/msg}) = 425\,\text{messages}.$$

Two mechanisms trigger the delayed discarding of new work. Both require protocol-specific knowledge and therefore involve delegation to applications:

1. Each message is timestamped upon arrival. Before a message on the ingress queue is processed, its timestamp is checked to determine how long it was queued. If the answer is greater than some protocol-specific threshold, the message is discarded in the knowledge that the request has timed out, or under the assumption that the user has given up or retried.

This strategy has the curious effect of favoring more recent requests over older ones. Users who have been waiting for a short time are served first, at the expense of users who have been waiting for longer and whose requests are discarded. The rationale for this is twofold. First, many users in the latter group will have given up or retried. Second, providing acceptable service to *some* users is better than providing unacceptable service to *all* users. In some cases, a LIFO queue implements this strategy, such that the most recent request is processed first. This postpones the need to discard stale requests, which allows the system to focus on work that it will accept.

2. When a protocol defines a message for explicitly cancelling a request, pairing requests and cancellations allows both messages to be discarded when a cancellation arrives. Because the pair of messages equates to a no-op, this strategy eliminates futile work. It would be used, for example, to discard the offhook–onhook message pair that results when a user gives up waiting for dial tone.

   To implement this strategy, messages must be associated with processing contexts as they arrive. When a message representing new work arrives, a state machine is created to handle the work. The message is queued on the state machine, and the state machine is placed on the ingress queue. When a message representing progress work arrives, the state machine that will handle it is located. Again, the message is queued on that state machine, and the state machine is placed on the progress queue *unless it is still on the ingress queue*. In the latter case, the initial message has yet to be processed. This is how a request-cancellation message pair is detected and discarded, along with the state machine that would have processed it.

   Under this strategy, it may not be appropriate to discard user requests that have been queued for a long time. Requests that are *known* to have timed out should be discarded, but discarding user requests for which explicit cancellations or retries can be identified is undesirable because it penalizes users who wait patiently instead of continually resubmitting their requests. However, the length of the ingress queue must be increased to reward patient users. Some telephone switches, for example, allow as many as 1000 messages on the ingress queue where users wait for dial tone, but still process them in LIFO order for the reasons discussed above.

In a stateless system, all work is new work. A stateless system therefore discards work when the length of its work queue reaches some threshold. It begins by discarding every *n*th message but lowers the value of n if the work queue continues to grow. Appropriate values of n for various queue lengths can be determined using queuing theory.

To guard against incoming message floods, many extreme systems disable I/O interrupts after a certain number of messages arrive within some interval. A LEAKY BUCKET COUNTER can trigger this action, or it can simply be triggered by measuring the time spent at I/O level. The latter approach effectively places the I/O ISR in a scheduler faction that receives a certain percentage of CPU time. Although disabling I/O interrupts will protect against message floods, it also results in the loss of progress work if the I/O buffer overflows because the CPU is not keeping up with incoming messages. The system must therefore reenable I/O interrupts regularly, perhaps once per second.

10.3 IGNORE BABBLING IDIOTS

A **babbling idiot** is a device or interface that is generating messages at an unacceptable or unexpected rate. An extreme system must protect itself from this kind of behavior because, whether it is intentional or not, it can have the effect of a denial of service attack.

Babbling occurs for a number of reasons:

- *Faulty hardware*. If a link is cut or suffers some other type of fault, it may generate a stream of garbage that software scanning the link translates into a flood of messages.
- *Faulty software*. A bug in application software can cause it to generate a flood of messages.
- *User actions*. A user can generate a flood of messages from a device by, for example, rapidly pressing buttons on a telephone set. Similarly, some users do not patiently wait for dial tone, but instead rattle the switch hook vigorously when the system is overloaded.

Critical components implement babbler detection on less critical components. The reason is that a babbler's messages are initially discarded, and a babbler is eventually reset or removed from service if its behavior persists.

A LEAKY BUCKET COUNTER can detect a babbling idiot. The counter is configured to accept up to n messages in t seconds, where the values of n and t are based on the type of device or interface that could babble. The values are chosen so that, under normal operating conditions, messages from the interface are accepted. When a message arrives, the counter is decremented. If t seconds have elapsed since the counter was last reset, it is again reset to n.

If the counter drops to zero, the interface is assumed to be babbling. Messages from the interface are discarded and corrective action is taken. On the first occurrence of babbling, the interface might be left in service, but its messages would be discarded until the counter was again reset to n. A second leaky bucket counter that tracks the number of babbler incidents within a given interval can detect a recurrence of babbling.

On a second occurrence of babbling within a prescribed interval, a hardware interface might be removed from service and subjected to a diagnostic. If the diagnostic passes, the interface returns to service. If it fails, the interface remains out of service until reenabled by an administrative command. In the case of a software interface or a user device, a reset message might be sent to it instead.

On a third occurrence of babbling, the interface might be removed from service until reenabled by an administrative command.

10.4 THROTTLE NEW WORK

The techniques discussed so far – handling progress work before new work, discarding new work, and ignoring babbling idiots – are examples of reactive overload controls. A processor can implement reactive controls without the cooperation of other processors.

Cooperating processors, however, can prevent overload situations from occurring. They can do so by throttling new work flowing from one processor to another. Note that progress work is never throttled, because a processor must never get into a situation in which it has taken on more work than it can handle.

The best way to throttle new work is with a credit scheme. Under this approach, a processor (a service node, for example) hands out credits to the processors (access nodes) that send it new work. If the service node has time for new work, it sends a message to each access node, telling it that it may send the service node some number of messages that represent new work.

Each access node uses a LEAKY BUCKET COUNTER to maintain the count of credits received from the service node. When the access node receives credits from the service node, it adds them to the counter while ensuring that the counter never exceeds a predetermined limit. When the access node sends new work to the service node, it decrements the counter. If the counter drops to zero, the access node queues or rejects new work until more credits arrive to replenish the counter. Note that, if an access node queues new work rather than discarding it, and if it also implements its own reactive overload controls, it will queue new work for both input (to itself) and output (to the service node).

When the service node reaches peak capacity, it simply stops giving out credits. It does not have to tell access nodes that it is extremely busy. Messages that hand out credits are the equivalent of 'The service node is running OK'. When the service node is too busy, the last thing it wants to do is spend time informing access nodes of this fact.

How does the service node know how many credits to hand out? In part, the answer depends on the frequency with which it gives out credits. Let's say that it gives out credits once per second. The service node must then know, either as the result of modeling or observation, how much work it can expect to complete within one second. It is not enough to simply accept new work; time must also be allotted for handling the average number of progress messages that will later arrive as the result of accepting the work. If the service node can handle 100 sessions per second, and it is serving 20 access nodes, it might give out 10 credits to each one, under the assumption that not all access nodes will initiate a full 10 sessions per second.

When the service node begins to approach an overload situation, it reduces the number of credits that it hands out. When it is running at full capacity, it stops handing out credits altogether. Then, as it begins to catch up with its backlog of new work, it gradually increases the number of credits to each access node.

How does the service node detect an overload situation? There are a number of detection mechanisms, all of which have proven successful in various systems:

- Monitor the length of the ingress queue (the queue of new work). If the length of this queue is increasing monotically, the system is in overload. An upper limit on the queue length usually serves to detect overload.
- Monitor the queue delay experienced by work items on the ingress queue. The system is in overload if the delay reflects an unacceptable response time.
- Monitor idle time (the time spent in the idler thread). Under PROPORTIONAL SCHEDULING, it is better to monitor faction occupancy (the percentage of its allotted time that a scheduler faction is using).

In each approach, observing the system under load determines the number of credits to hand out at various thresholds. Modeling and queuing theory can also play a role in the determination.

To throttle new work using a credit scheme, certain preconditions must be met:

- The processors need to cooperate. Although cooperation is possible among the processors that comprise an extreme system, it cannot be extended to processors that are external to the system in the absence of standards that specify the expected behavior.

- Throttled processors must be able to distinguish new work from progress work.
- Only a limited number of processors can be throttled. If there are too many of them, the number of credits that can safely be given to each will be too low to accommodate what amounts to a random distribution of newly arriving work.

When a sink (destination) receives work from multiple sources, it must hand out the credits, but when a sink receives work from a single source, the source can impose the credit scheme on itself. For the degenerate case of a single source, this solution is simpler.

10.5 SUMMARY

- A system needs overload controls to prevent thrashing and crashing when it receives too much work.
- In stateless systems, overload controls typically discard every *n*th request.
- In stateful systems, overload controls distinguish ingress work (a message that creates a new session) from progress work (a message to an existing session). To service the ingress work queue, the progress work queue must be empty.
- Two additional work queues, an immediate queue and an egress queue, may be required. The immediate queue has the highest priority, and the egress queue only has higher priority than the ingress queue.
- When you dequeue new work, discard it if it has timed out as the result of sitting in the queue for too long.
- Match up requests and cancellations in the ingress queue so that you can discard both of them.
- Match up retransmitted requests in the ingress queue so that you can discard duplicates.
- If the ingress queue becomes so long that it could prevent resources from being available for progress work, discard new work immediately. Depending on the protocol involved, you might discard the new request (LIFO order) or accept it and discard the oldest request (FIFO order).
- Use LEAKY BUCKET COUNTERS to detect babbling idiots. Abort their work, reset them, or deny them further service.
- To prevent work sources from sending you too much new work in the first place, send them credits for new work when you are able to accept it.

11

Failover

Extreme systems often configure processors which can take over the work of processors that fail as the result of hardware or software faults. Such redundancy avoids total or partial outages, or reduces their duration.

This chapter primarily describes redundancy strategies which occur in the types of stateful systems that are the primary focus of this book. For a broader survey of redundancy strategies, see [SAR02].

11.1 REDUNDANCY TECHNIQUES

Let's start by defining some terms that apply to redundancy strategies:

- An **active processor** is one that is currently handling work.
- A **standby processor** can take over work when an active processor fails.
- **Failover** occurs when a standby processor takes over the work of an active processor. The standby becomes the active processor, and the active processor becomes a standby after it recovers from its failure.

There are many ways to configure active and standby processors (see Figure 11.1). We will now describe them, using the following criteria to assess their strengths and weaknesses:

- *Hardware cost.* How many processors are needed? Is any custom hardware development necessary?

Robust Communications Software G. Utas
© 2005 John Wiley & Sons, Ltd ISBN: 0-470-85434-0 (HB)

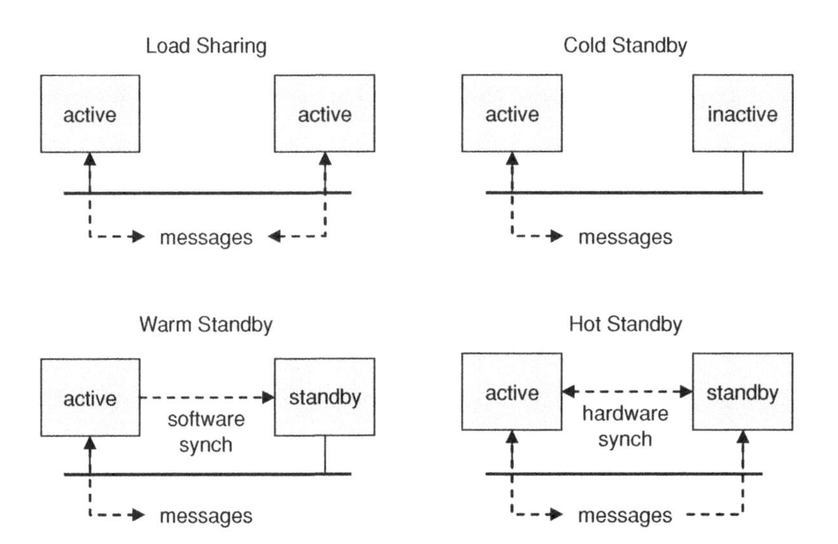

Figure 11.1 Failover techniques. Load Sharing splits the workload among more than one processor. Cold Standby allows a backup processor to take over, but it must first be configured. Warm Standby and Hot Standby provide synchronization so that a standby processor can immediately take over.

- *Hardware failure transparency*. What is the impact of a hardware failure that causes a failover?
- *Software cost*. Does application software require modification to support failover?
- *Software failure transparency*. What is the impact of a software error that causes an outage in the active processor, and thus a failover?
- *Recovery time*. If a failover occurs, how quickly does the system recover?
- *Capacity impact*. How much is the system's capacity reduced by supporting this form of failover?

11.1.1 Load Sharing

Under LOAD SHARING, a group of independent, active processors runs in parallel, sharing the workload. If a processor fails, the other processors take over its work. Some form of load balancing, such as a round-robin scheme, evenly distributes work among the processor group.

Load sharing has the following characteristics:

- *Hardware cost*: $n + m$ processors, using off-the-shelf hardware. The values of n and m are chosen so that n processors can handle the peak workload if m processors fail.

- *Hardware failure transparency*: All work in a failed processor is lost.
- *Software cost*: Transparent to applications.
- *Software failure transparency*: All work in a failed processor is lost.
- *Recovery time*: Excellent. The survivors immediately handle all of the work.
- *Capacity impact*: Each processor runs at $n/(n+m)$ of its maximum capacity to leave headroom in case m processors fail.

Stateless servers that implement simple request-response protocols are often configured in a load sharing arrangement. An example is a name server, such as a DNS server. Stateless servers usually perform database lookups. In a loading sharing arrangement, the database is replicated in all processors. Load sharing between stateless servers is easy because all messages to the servers can be routed independently.

11.1.2 Cold Standby

Under COLD STANDBY, an *inactive* standby processor takes over when an active processor fails. This is also known as '$n+m$ sparing' because there are n active processors and m standbys. The values of n and m are chosen so that n active processors can handle all the work, with m standbys ensuring the required level of availability when some of these processors fail.

Cold standby has the following characteristics:

- *Hardware cost*: $n+m$ processors, using off-the-shelf hardware.
- *Hardware failure transparency*: All work in a failed processor is lost.
- *Software cost*: Transparent to applications.
- *Software failure transparency*: All work in a failed processor is lost.
- *Recovery time*: Often slow because a standby processor must download data before it can assume the role of a failed processor. The recovery time improves significantly in an $n+n$ configuration that pairs active and standby processors and preloads each standby with the same configuration data as its active counterpart.
- *Capacity impact*: None.

Cold standby is often used in a large pool of homogeneous processors, when the failure of any one of the processors only causes a partial outage. Service nodes (see Section 2.2) often run in a cold standby arrangement. The reason is that LOAD SHARING among stateful servers is difficult because the server that handles the initial message in a connection-oriented protocol must receive all subsequent messages in the resulting session. Furthermore, stateful servers usually run services on behalf of subscribers. For each subscriber, a profile determines if the subscriber can access a particular service and, if so, how it should behave. The aggregate size of these profiles is typically so large that replicating them in a load-sharing configuration would be too expensive. Consequently, the system assigns each subscriber to a specific service node and only reassigns the subscriber to another service node if the original one fails.

11.1.3 Warm Standby

Under WARM STANDBY, two processors run independently, and the active one handles all external I/O. The active processor regularly sends information to the standby processor so that objects in the standby remain closely synchronized with those in the active processor. The synchronization need not be perfect, but it must be sufficient to allow the standby to preserve most of the work if the active processor fails. The synchronization process is known as **checkpointing**.

Warm standby has the following characteristics:

- *Hardware cost*: $2n$ processors, using off-the-shelf hardware.
- *Hardware failure transparency*: Work in the failed processor survives to the extent that objects in the standby processor are up to date.
- *Software cost*: The active processor must checkpoint application objects to its standby. Different ways to do this are discussed later in this chapter; some of them involve adding software to each application that needs to mirror objects in the standby processor.
- *Software failure transparency*: Again, work in the failed processor survives to the extent that objects in the standby processor are up to date. If the synchronization has also placed the standby processor in a faulty state, an outage occurs. The risk of this is reduced if checkpointing only occurs at the end of each transaction.
- *Recovery time*: Good. Failover should occur within one second.
- *Capacity impact*: The time required for checkpointing reduces capacity.

Warm standby is often used in stateful servers when an outage is undesirable. Telephone switches, for example, often replicate a call's objects in a standby processor, but only after answer occurs. If the active processor then fails, all conversations continue. However, any call in a dialing or ringing state is dropped, and the subscriber must reinitiate it.

11.1.4 Hot Standby

Under Hot Standby, two processors (each with its own memory) run in synch at the hardware level, simultaneously executing the same instructions. Both processors receive messages, but only the active processor sends external messages. Custom hardware monitors the behavior of both processors. If it detects a mismatch (perhaps in their program counters), it forces each one to run a diagnostic to determine if it is healthy. The healthy processor assumes (or retains) control, and the faulty processor is removed from service until its fault is corrected.

Hot standby has the following characteristics:

- *Hardware cost*: 2*n* processors, with custom hardware to couple each pair.
- *Hardware failure transparency*: All work in the failed processor survives because it is fully replicated on the standby processor.
- *Software cost*: Transparent to applications – as far as *failover* is concerned. However, the checkpointing of application objects, in the same way as under Warm Standby, may nonetheless be required for reasons discussed in Section 12.2.
- *Software failure transparency*: Synchronized insanity. Both processors run the same software in lock step. They therefore fail together, causing an outage.
- *Recovery time*: Excellent for hardware failures. Poor for software failures because a restart occurs.
- *Capacity impact*: The checkpointing hardware slows down each processor to some extent.

Hot standby is sometimes used in high-end, stateful servers. A telephone switch might implement a centralized function in two processors that run in hot standby. Generally, however, Warm Standby is preferable because it requires no custom hardware and is less vulnerable to software failures. In systems of even modest complexity, software failures cause more failovers than hardware failures, in which case hot standby's synchronized software insanity is better avoided.

11.2 CHECKPOINTING TECHNIQUES

Under Warm Standby, checkpointing is required so that changes to application objects in the active processor are reflected in the standby processor. This section outlines techniques for implementing checkpointing. For an overview of checkpointing, see [HAN03]

During normal operation, an active processor sends updates to its standby as changes occur. In this case, application objects build checkpoint messages when they are created, modified, or deleted. The active processor sends the messages at the end of each transaction.

Checkpointing is also required when a standby returns to service after an outage. Here, the active processor must send the standby processor a large number of messages to bring it up to date. In this case, some form of catch-up function must prompt applications to perform bulk checkpointing. The active processor cannot wait for an object's data to change before checkpointing it to the standby.

In Section 12.2, we will describe how bulk checkpointing from one software release to another can preserve work during a software upgrade. Because different software releases have different memory layouts, a technique that supports checkpointing between them cannot require identical memory layouts. A technique that requires identical layouts can only be used when the active and standby processors are running the same release.

11.2.1 Application Checkpointing

APPLICATION CHECKPOINTING adds checkpointing software to applications on an as-needed basis. Although this solution is *ad hoc*, it is more efficient than generic approaches.

Each application defines the contents of its checkpoint messages and, during normal operation, decides when to send them. An application also registers a callback that is invoked to perform bulk checkpointing. On the standby processor, the application provides software to receive its checkpoint messages and perform the associated updates.

11.2.2 Object Checkpointing

Although APPLICATION CHECKPOINTING is relatively efficient, it is also a burden to application developers. OBJECT CHECKPOINTING provides a framework to remove some of the burden.

1. During a transaction, objects add themselves to a checkpoint registry when their data changes.
2. At the end of the transaction, the registry invokes a `Pack` function on each object to add it to a checkpoint message. Each object can be encoded as a parameter in a TLV MESSAGE. The parameter type corresponds to the object's class, and the parameter value

(a `struct`) contains the subset of data that the `Pack` function saves to serialize the object.

In a checkpoint message, an object identifier must replace a pointer member unless both processors have identical memory layouts. If a registry tracks all objects in a class, as discussed in Section 4.2.3, this identifier can be an index into the registry's array. If an OBJECT POOL allocates all objects in a class, the identifier can be an index into the pool's array of blocks. Because a large registry or pool typically uses a two-dimensional array, the identifier usually combines an upper and lower array index.

3. The checkpoint registry sends the checkpoint message to the standby processor, where it is demultiplexed. During demultiplexing, an `Unpack` function is invoked on a singleton that corresponds to each parameter's class. This singleton updates the mirrored instance of the object in the standby. The `Unpack` function also regenerates pointer members by mapping object identifiers to pointers.

The `Object` class can define virtual `Pack` and `Unpack` functions. The default versions do nothing, and objects that support checkpointing override.

To perform bulk checkpointing, the framework uses a registry of application singletons. Each singleton provides an iterator for cycling through the objects that require checkpointing. The framework again uses the `Pack` and `Unpack` functions while building and demultiplexing the checkpoint messages.

11.2.3 Memory Checkpointing

Although OBJECT CHECKPOINTING provides a checkpointing framework, it still involves writing a lot of object-specific code. A customized cache can eliminate such code, but it assumes identical memory layouts in the active and standby processors. Memory write operations occur in a cache that holds memory pages dirtied by write operations. The cache is customized to generate a list of the modified pages. At the end of each transaction, the modified pages are locally committed and also checkpointed to the standby processor.

11.2.4 Virtual Synchrony

The checkpointing techniques outlined thus far share some drawbacks. First, their processing overheads degrade capacity. These overheads arise because the standby processor does nothing during a transaction. Instead, the active processor performs checkpointing at the end of each transaction, when it assembles and sends a considerable amount of data to the standby. Second, MEMORY CHECKPOINTING assumes an identical memory layout in both processors. APPLICATION CHECKPOINTING and OBJECT CHECKPOINTING need not do so, but they must then regenerate pointers, which adds to their cost and complexity.

HOT STANDBY reduces checkpointing overhead because the active and standby processors perform the same work in parallel. VIRTUAL SYNCHRONY achieves the same result by approximating hot standby in software. It presents each incoming message to both the active and standby processors, which run in parallel. Outputs (messages) from the two processors are compared. The comparison may be performed by the active processor or by another processor that acts as a front-end to both the active and standby processors. If the results match, the processors are assumed to be in synch, so the outgoing messages from the active processor are sent. If the results differ, the processors have gotten out of synch, so one of them continues in service while the other one is reset and resynchronized. If both processors respond, the active one is deemed correct. A processing error, however, usually results in a timeout or an exception, in which case the processor that responds is deemed correct.

On the surface, virtual synchrony is an elegant solution. Its processing overhead consists of a post-transaction comparison, which is more efficient than the post-transaction checkpointing of the other approaches. And because the processors handle transactions independently, their memory layouts need not be identical.

Unfortunately, things are not that simple. The following issues arise:

- Although the processors synchronize on external events (the incoming messages), internal events can cause them to lose synch. The expiration of timers is a good example. Because the processors do not synchronize on each CPU instruction, they do not start timers at precisely the same time. A timer can therefore expire in one processor but not the other, causing their behavior to diverge. To overcome this problem, timers must somehow be synchronized. One way to do this is to run all timers in the front-end processor, if it exists, or in the active processor. In other words, all inputs, whether internal or external, must come from a common source.
- A processor that uses the overload control FINISH WHAT YOU START (see Section 10.1) does not process inputs in FIFO order. When

a message represents new work, a considerable delay can occur between the time at which it arrives and the time at which it is processed. This does not create a problem *per se*, but it means that a series of messages in the output stream is seldom the result of processing the last message received in the input stream.

- Because both processors execute the same software in parallel, they can experience the same type of synchronized software faults that plague Hot Standby. However, this outcome is not certain because some software faults arise from race conditions. The processors do not synchronize on each instruction as they do under Hot Standby, so one of them may avoid a race condition which the other one encounters.
- To resynchronize the two processors after a failure resets one of them, some form of bulk checkpointing is required. It will be based on one of the previous approaches, which again leads to the need for identical memory layouts or mapping between pointers and object identifiers.

Virtual synchrony is a popular research topic. At the time of writing, a Google search on the phrase returned over 4000 hits. Our treatment of it is focused and pragmatic in that it only seeks to approximate Hot Standby between two processors in a tightly coupled system. Research papers, however, take a much broader view. They allow an arbitrary number of processes and therefore define protocols for managing process groups. Members of a group run in virtual synchrony; new processes can join the group; failed processes are removed from the group. Processes can be geographically dispersed, so issues of message loss and reordering require attention. This form of virtual synchrony often assumes that a process joining the group can catch up with the others by replaying previous transactions, which are saved. This technique is more suited to transaction processing than session processing, both because of its overhead and because of the collaborative role that session participants – now absent – play in connection-oriented protocols. The absence of these participants makes it difficult to replay transactions.

11.3 ADDRESSING PROCESSORS

The processors in an active–standby configuration require both physical and logical message addresses. Applications want their messages to go to the active processor, but they do not want to worry about which processor is actually active. System maintenance software, on the other hand, needs to address processors explicitly so that, for example, it can tell the standby processor to become active. Three IP addresses are therefore required:

1. A physical address for processor 0, used by software that
   - wants to address this specific processor, or
   - knows which processor is active and which is standby, and wants to address the standby processor.
2. A physical address for processor 1, used in the same way as the address of processor 0.
3. A logical address for the active processor. It resolves to the physical address of the active processor and is used by the vast majority of software, which is only interested in addressing the active processor.

11.4 INTERACTING WITH ESCALATING RESTARTS

When a node supports both failover and ESCALATING RESTARTS, it must decide which approach to use when recovering from a failure. Failover to the standby processor is the obvious choice when the active processor suffers a hardware failure. In the case of a software failure, however, restarting the active processor may be preferable. There are two considerations:

1. Which approach will be faster?
2. Which approach will have less of a service impact?

Because the processors in a HOT STANDBY configuration run in synch, any software failure will occur in both of them. A failover will not fix this problem, so a restart is the logical choice. A failover may only help if the active unit fails to reboot. In this case, we'd better hope that we're dealing with a hardware problem. For hot standby, the likely escalation sequence is therefore:

1. Warm restart on processor A (the one that failed).
2. Cold restart on processor A.
3. Reload restart on processor A.
4. Reboot of processor A.
5. Failover to processor B, followed by an immediate reboot of processor B.

During steps 1 to 4, processor B remains on standby.

WARM STANDBY usually confines a software failure to one processor, in which case failover is more appealing. Initially, the choice is between a warm restart and a failover. A warm restart is preferable because it has less service impact. It loses fewer messages because it does not update the node's logical message address. It also preserves work better, assuming that checkpointing does not include transient states. Its main drawback is that, because it kills and recreates threads, it is somewhat slower than a failover.

If the warm restart fails and would escalate to a cold restart, failover is more appropriate. It will be faster than a cold restart and, more importantly, it has less of a service impact because it preserves checkpointed work.

For warm standby, the likely escalation sequence is therefore:

1. Warm restart on processor A (the one that failed).
2. Failover to processor B.
3. Warm restart on processor B.
4. Cold restart on processor B.
5. Reload restart on processor B.
6. Reboot of processor B.
7. Failover to processor A, followed by an immediate reboot of processor A.

During step 1, processor B remains on standby. During steps 2 to 6, processor A remains on standby.

11.5 SUMMARY

- Failover improves availability by allowing another processor to take over the work of a processor that fails.
- LOAD SHARING splits work among multiple processors that implement stateless transactions. If one processor fails, the others continue to provide service.
- COLD STANDBY configures a spare, inactive processor to take over the work of a failed processor.
- WARM STANDBY runs two processors in an active–standby configuration. The active processor sends checkpoint messages to the standby to keep it in synch. If the active processor fails, the standby takes over and preserves the checkpointed work.
- HOT STANDBY runs two processors in an active–standby configuration. The processors synchronize on each CPU instruction. If the active processor fails because of a hardware fault, the standby seamlessly takes over its work. But if it fails because of a software fault, a restart occurs because both processors are in exactly the same state.

- Application Checkpointing uses application-specific messages to maintain synchronization in a Warm Standby configuration.
- Object Checkpointing provides a framework for Application Checkpointing.
- Memory Checkpointing provides synchronization between two processors with identical memory layouts by sending modified pages from the active to the standby processor.
- Virtual Synchrony provides synchronization by sending each input to both processors, which run in parallel.
- If a processor in a Warm Standby configuration fails, bulk check-pointing is required to bring it back into synch with the active processor.
- Failover and Escalating Restarts are two different ways to recover from serious software errors. In Hot Standby, a full sequence of restarts on the active processor should precede a failover. In Warm Standby, failover occurs earlier, when a warm restart fails to keep the active processor in service.

12

Software Installation

When an extreme system updates its software, its goal is to avoid any service disruption. A procedure for installing new software is **hitless** if it satisfies this requirement. Customers of extreme systems want to avoid outages; it makes little difference to them whether an outage is unplanned (the result of a hardware or software failure) or planned (an unavoidable effect of performing a software update).

There are two types of software updates. The first is a **patch**, which delivers bug fixes. The second is an **upgrade**, which delivers new capabilities. A patch is often referred to as a 'dot release' because many systems name it using a convention like 'Release 3.1'. An upgrade is then a 'dot-zero' release, such as 'Release 4.0'.

In many systems, the procedures for installing a patch or an upgrade are identical: halt the system and reboot it with a new load. The only distinction between a patch and an upgrade is that a patch fixes bugs, whereas an upgrade implements new capabilities. Both of them, however, cause an outage.

Some systems avoid a total outage by replacing individual processes. In Section 7.2.8, however, we discussed why an extreme system limits its use of processes and why killing and restarting a process typically disrupts service in any case.

An extreme system must therefore seek less disruptive ways to update its software. In this chapter, we will discuss how to install patches and upgrades in a hitless manner. Hitless patching is quite feasible, although it involves some effort and is subject to certain restrictions. Hitless upgrades are more challenging and typically minimize, rather than totally avoid, service disruptions.

Robust Communications Software G. Utas

12.1 HITLESS PATCHING

The patching procedure for an extreme system takes advantage of the fact that most bugs arise from logic errors in application code. These bugs can be fixed solely by modifying object code (that is, .cpp files). Indeed, HITLESS PATCHING requires that all function signatures and data structures remain the same, but this restriction seldom poses a problem given that most patches only need to fix logic errors.

12.1.1 Function Patching

The approach that underlies hitless patching is simple. A portion of memory is reserved for object code that implements patches. A new version of a function is written to fix a bug, and the code is compiled to obtain new object code. The new object code is loaded into the patch area, and the old version of the function is modified to immediately call the new version of the function. The new version either returns through the old function or is modified to unwind the stack to the original calling function.

Note that the old version of the function is not removed. This serves a number of purposes:

- If a thread was scheduled out in the middle of the old function, it can safely resume execution. However, this rarely occurs under our extreme programming model because threads only block in a few places and use COOPERATIVE SCHEDULING to minimize preemption.
- Software that invokes the old function need not be modified to call the new function. All calls continue to go through the old function.
- The old function remains available. It can therefore be quickly reinstated if the new version causes unforeseen problems.

By restricting itself to replacing object code, this technique installs patches in a totally hitless manner. Implementing it is relatively straightforward, even if the details are something of a black art. The basic requirement is an incremental loader that

- inserts the object code for a new function;
- modifies the old function to invoke its replacement;
- updates symbols so that debug tools continue to work correctly, and
- invalidates the instruction cache in case it contains the old function.

12.1.2 Class Patching

Although the primary purpose of patches is to fix bugs, they sometimes deliver new capabilities. This situation arises when a customer simply cannot wait for the next software release to obtain a critical new capability. If the product supports hitless patching, the customer then asks for the capability to be delivered as a patch.

Delivering a new capability as a patch is challenging because a new capability usually requires **NUC** (non-upward compatible) changes. A NUC change is one that forces the recompilation of an interface's users. Most interface changes (in .h files) are NUC changes. Examples include

- adding member data;
- changing a function's arguments;
- adding or overriding a member function;
- inheriting from a different class or an additional class;
- changing the value of a compile-time constant.

There are exceptions in which some of these are not NUC changes, but we will not discuss them here because it would embroil us with compiler details.

A patch cannot make a NUC change because this could involve replacing the object code for hundreds, or even thousands, of functions. Worse yet, a NUC change in member data would cause the new version of a function to read data from incorrect offsets in objects that were constructed before the patch was installed. Consequently, patches can only modify implementations (.cpp files).

Fortunately, however, there is a rather simple way to avoid most NUC changes. Although it is not a complete solution, in practice it accommodates most patches that would otherwise require NUC changes. It does so by providing back-door mechanisms that effectively allow a class to add member data and functions. The definition of the `Object` class in Section 4.1 provided these mechanisms, but we have postponed discussing them until now.

The root class `Object`, from which most classes derive, defines `patchData_`, a pointer to any object. If an object needs more member data, it defines the data in a new object and references it through `patchData_`. Defining a new class for this object is not a NUC change.

The `Object` class also defines a virtual `Patch` function. It takes two arguments: an `int` used in a `switch` statement that selects the code associated with a patch, and a `void` pointer that is cast as whatever `struct` carries the actual arguments required by a specific usage of the `Patch` method. To preserve inheritance, each class must override `Patch` by invoking its superclass `Patch` function. The default code looks like this:

```
void Object::Patch(int selector, void *arguments) { }

void SomeObject::Patch(int selector, void *arguments)
{
   Object::Patch(selector, arguments);
}
```

A new function needs a selector that identifies it and a `struct` that defines its arguments and any values that it returns. For example:

```
const int NewFunctionId = 10;

struct NewFunctionArgs
{
   int arg1;
   int arg2;
   int result;
};

void SomeObject::Patch(int selector, void *arguments)
{
   switch(selector)
   {
   case NewFunctionId:
      NewFunctionArgs *args =
         (NewFunctionArgs*) arguments;
      //
      // code for new function, perhaps calling
      // superclass Patch
      //
      return;
   }
   Object::Patch(selector, arguments); // for other
                                       // selectors
}

int NewFunction(SomeObject &obj, int arg1, int arg2)
{
   NewFunctionArgs args;
   args.arg1 = arg1;
   args.arg2 = arg2;
   obj.Patch(NewFunctionId, &args);
   return args.result;
}
```

Member data and functions can now be added without NUC changes. Designers can use the same mechanism during testing, as it is much faster to patch code than it is to rebuild a large system. The `Patch` function, however, is never used in code submitted to the development stream, except for the mandatory invocation of superclass `Patch`.

12.1.3 Managing Patches

What we have described thus far is only a starting point. An extreme system must also provide a comprehensive patch administration system that supports the following types of capabilities:

- Allow administrators to download patches from a web site or FTP server.
- Refuse to install a patch if any prerequisite patches are not yet installed.
- Remove any superseded patches when installing a patch.
- In a HOT STANDBY or WARM STANDBY configuration, install the patch on both processors. If the standby processor is out of service, install the patch on it as soon as it returns to service.
- Allow a patch to be backed out if it causes problems.
- Allow a patch to be committed once it has been proven, so that it cannot be inadvertently removed. After a reboot, reinstall committed patches. However, do not reinstall any uncommitted patch, because it might have caused the reboot.
- Provide an inventory of all patches that have been installed.

Most of the work actually lies in implementing the patch administration system rather than the incremental loader itself. Without a patch administration system, the risk of procedural errors is too great, and some of these will cause outages that could otherwise have been avoided.

The software development process must also take patching into account:

- Developers must check patches into the software library.
- Patches are only checked into branches that have been released to the field.
- Branches in which development is still underway must also resolve bugs fixed by patches. In some cases, the patch's solution is acceptable. In others, a better solution must be found because the purpose of patches is to fix problems as fast as possible, with little regard for architectural principles.

- If a function has been patched, it should not be repatched. The new version of the function should incorporate both fixes, superseding the original patch.
- Because an inlined function is cloned into its callers, it cannot be patched. Developers must consider this when using inlines.
- Once a software release contains hundreds of patches, patch administration becomes unwieldy and error prone. At that point, customers may prefer to install a true dot release that incorporates all of the patches.

12.2 HITLESS UPGRADE

A full software upgrade delivers new capabilities. Unlike a patch release, a new release contains NUC changes that inevitably result from implementing its new capabilities.

One way to minimize the disruption associated with a full software upgrade is to build upon the WARM STANDBY version of failover (see Section 11.1.3). A HITLESS UPGRADE follows these steps:

1. Take CPU1 (the standby processor) out of service.
2. Insert the new software load in CPU1.
3. Load CPU1 with configuration data.
4. Put CPU1 back into service.
5. Wait for CPU0 (the active processor) to resynchronize CPU1.
6. Manually initiate a failover to CPU1.
7. Take CPU0 (now the standby processor) out of service.
8. Insert the new software load in CPU0.
9. Load CPU0 with configuration data.
10. Put CPU0 back into service.
11. Wait for CPU1 (now the active processor) to resynchronize CPU0.

Because this approach involves a failover, it is not completely hitless. However, it dramatically reduces the impact and duration of outages caused by software upgrades. During the upgrade procedure, the system always has an active processor to perform work. However, it runs without a standby processor except during step 6.

Steps 3 and 9, which load the new release with configuration data, present a challenge. The configuration data already exists on a control node's disk, but in the format of the previous release. If, as is invariably the case, the new release added or deleted fields in some of this data, the data must be reformatted before it can be loaded into the new release. Either the node that is sending or the node that is receiving the data can perform this OBJECT REFORMATTING. Another option is to prepare a BINARY DATABASE in advance, as described in Section 8.7.7, and load it directly.

In step 5, CPU1 is running the new release, whereas CPU0 is running the old release. This presents a challenge to resynchronization software because the new release will have made NUC changes to some classes. In Section 11.2, we discussed three ways to provide resynchronization: APPLICATION CHECKPOINTING, OBJECT CHECKPOINTING, and MEMORY CHECKPOINTING. We can now see that memory checkpointing cannot support hitless upgrades, because the old and new software releases will not have identical memory layouts. Even under application or object checkpointing, object reformatting is required. For example, the `Unpack` function in the new release must be capable of creating objects from the old release. Such complexity is why only the most extreme of extreme systems support hitless upgrades.

After step 6, the new release (in CPU1) becomes active. Step 7, and especially step 8, are therefore delayed until the new release has run for some time without embarrassment. Indeed, the upgrade process must support rollback procedures at each step, so that sanity can be restored if something goes wrong.

The procedure for performing a hitless upgrade under HOT STANDBY is identical to the one just outlined. A hot standby system must therefore support the WARM STANDBY form of resynchronization (that is, bulk checkpointing) to offer hitless upgrades.

12.3 ROLLING UPGRADE

An extreme system must upgrade its processors one at a time, a procedure known as a ROLLING UPGRADE. The reasons for this, and its ramifications, were discussed in Section 6.2.4. Even under a HITLESS UPGRADE, a rolling upgrade is standard practice because operators are reluctant to suddenly switch an entire system to an unproven, new release.

In a rolling upgrade, the main question is the order in which to upgrade the processors. A system based on the reference model of Section 2.2 usually upgrades its nodes in the order shown in Figure 12.1:

1. Administration nodes are upgraded first. This allows new configuration data, defined in the new release, to be provisioned ahead of time. Administration nodes must therefore be able to communicate with any node that is running the previous software release.

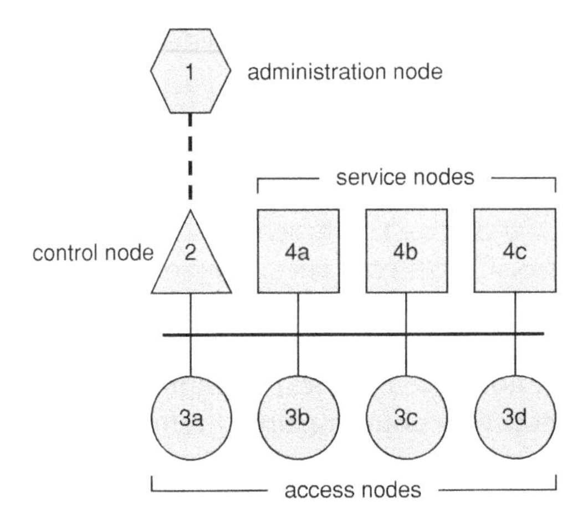

Figure 12.1 Rolling upgrade. A new software release is deployed one node at a time rather than in parallel. The administration node is upgraded first, the control node second, the access nodes third, and the service nodes last.

2. Control nodes are upgraded next. This prepares the way for a rolling upgrade of service and access nodes. A control node must therefore be able to communicate with service or access nodes that are still running the previous software release.
3. The software in access nodes is simpler than that in service nodes, so it is upgraded next. An access node must therefore be able to communicate with service nodes or other access nodes that are running the previous software release.
4. The software in service nodes is usually the most complex, so it is upgraded last. A service node must therefore be able to communicate with other service nodes that are running the previous software release.

A rolling upgrade raises the question of how long to maintain PROTOCOL BACKWARD COMPATIBILITY. Clearly, release n must support communication with release $n - 1$, but what about release $n - 2$ or even release $n - 3$? Once a product attains a certain level of functionality and maturity, customers may not wish to take every new release. Installing a new release always entails risk, and the customer may not even need the new capabilities that it offers. Consequently, a mature system often needs to support upgrades from release $n - 2$, or even $n - 3$, to release n.

12.4 HOW HITLESS DO YOU NEED TO BE?

Based on this chapter and the previous one, it should be clear that implementing hitless failover is difficult. VIRTUAL SYNCHRONY is the most hitless solution when both processors are in service, but it is difficult to implement and does not support the bulk checkpointing capability that is required when bringing a standby processor into service.

Either APPLICATION CHECKPOINTING or OBJECT CHECKPOINTING can implement bulk checkpointing, but they rarely provide totally hitless failovers. There are so many data changes that propagating all of them would seriously degrade capacity. Systems that perform session processing rarely checkpoint objects after each transaction. They usually checkpoint a session's objects only after it attains a stable state. The definition of *stable* is application specific but generally excludes transient states, such as sessions in a setup or disconnect phase.

Hitless failover (bulk checkpointing under HOT STANDBY or WARM STANDBY) is a prerequisite for a HITLESS UPGRADE, which can then be implemented using the procedure outlined in Section 12.2. However, scheduling software upgrades for off-peak times minimizes their impact. It may therefore be acceptable for them to cause a brief but total outage.

When a system supports hitless failover, customers nonetheless view unplanned failovers as unacceptable. A failover is always a major incident whose root cause must be found and fixed. Therefore, given the difficulty of implementing hitless failovers, and the fact that your efforts won't reduce your stress and workload anyway, there is a cruel but fair conclusion: Why bother? HITLESS PATCHING and ESCALATING RESTARTS should satisfy most customers. Only the most extreme of extreme systems need to provide hitless failovers and hitless upgrades.

12.5 SUMMARY

- An extreme system must be able to install new software without disrupting service to subscribers.
- An upgrade delivers new capabilities, whereas a patch delivers bug fixes.
- HITLESS PATCHING loads new object code for a modified function while a system remains in service. When the new code is in place, the previous object code is modified to jump to it.
- Patches do not permit NUC (non-upward compatible) changes to interfaces.
- The `Object` class defines a `patchData_` pointer and a `Patch` function to provide a back door mechanism for implementing NUC changes.

- A patch administration system allows customers to download, install, back out, and commit patches. It significantly reduces the risk of procedural errors.
- A Hitless Upgrade installs a new software release with minimal disruption. It takes the standby processor out of service and loads it with the new software and data. It then returns the standby to service, resynchs it using the bulk checkpointing capability of Warm Standby, and then forces a failover to the new release.
- A Rolling Upgrade upgrades processors to a new release serially instead of in parallel. The gradual introduction of a new release minimizes risk and the duration of outages. Protocol Backward Compatibility is a prerequisite for this strategy because there will be a lengthy interval when some processors are running the old release and others are running the new release.

13

System Operability

An extreme system must provide a number of capabilities that allow craftspeople to operate it smoothly. Craftspeople must be able to configure the system, observe its behavior, and issue commands to control its activities. When faults occur, the system should react to them autonomously so that it remains in service without human intervention. If this is impossible, it must notify craftspeople so that they can take corrective action. In the telecom domain, the terms operations, administration, maintenance, and provisioning (collectively abbreviated to **OAM&P**) describe these activities:

- *Operations* refers to craftspeople monitoring the system's health by analyzing information that the system provides about its behavior.
- *Maintenance* refers to actions taken to recover from faults. The system often performs maintenance actions autonomously, but craftspeople must perform them in some situations (to replace a faulty card, for example).
- *Provisioning* refers to craftspeople configuring the system by populating it with data (subscriber profiles, for example).
- The term *administration* is used in different ways. Sometimes it encompasses all of the above activities, as in *administration node*. This is the meaning used in this book, but in other contexts it refers to provisioning.

A lengthy book could be written about OAM&P, so this chapter only provides an overview that primarily deals with requirements. Besides space limitations, another reason for limiting our discussion of OAM&P is that it rarely introduces new extreme software

Robust Communications Software G. Utas

ISBN: 0-470-85434-0 (HB)

techniques. Its primary focus is operability, and its details depend on the types of application that a system supports. For further details on OAM&P, see [ADAM96], [HAN99a], [HAN02], and [HOH03].

Recall our extreme system reference model of Section 2.2. Craftspeople control and observe an extreme system through an administration node. The system itself consists of a control node and a number of service and access nodes. Commands issued on an administration node cause data to flow into the system, and the system sends data to the administration node to report on its activities.

The user interface that OAM&P software presents to craftspeople requires careful design because it quickly attains a state of inertia. There have been cases in which a firm redesigned its OAM&P interface after being told that it was not user friendly, only to learn that they could not deploy the improved system. In some cases, the reason was that craftspeople were already familiar with the old version and therefore objected to switching to the new, improved version. In other cases the reason was that the customer had written software to interface with the old version and did not want to change their software to support the new version.

13.1 ADMINISTRATION

This section discusses data that administration nodes download to the system (control, service, and access nodes) when craftspeople configure the system.

13.1.1 Configuration Parameters

CONFIGURATION PARAMETERS allow operators to engineer a system's size or control other aspects of its behavior. Previous chapters already mentioned the need to engineer software resources such as

1. the number of blocks in an OBJECT POOL,
2. the number of threads in a THREAD POOL, and
3. the size of the WRITE-PROTECTED MEMORY segment.

However, it is difficult determine appropriate values for such parameters without knowledge of the system's internal architecture. The system should determine the values itself, based on other configuration parameters that limit the operator to a black box, rather than a white box, view of the system. In a processor that controls telephone calls, for example, the above values might be determined by

1. the maximum number of simultaneous calls in the processor,
2. the maximum number of calls that the processor handles per second, and
3. the maximum number of subscribers served by the processor.

Even this, however, is unsatisfactory, because the first and last values in turn depend upon the percentage of time that a typical subscriber uses the telephone (occupancy) and the average duration of each call (holding time). The second value is system specific and must therefore be provided to the operator. What is ultimately required is a spreadsheet that calculates the appropriate values based on inputs that are familiar to the operator. This reduces the risk of parameters being set to inappropriate values. Nonetheless, the system should ensure that critical parameters are set to reasonable values. For example, it is desirable to enforce minimum values for the size of object pools, thread pools, and the size of the protected memory segment.

Some configuration parameters control the system's behavior rather than its software resources. The system usually provides default values for such parameters but allows them to be overridden if desired. For example, protocol timeouts are often configurable. The number of events and the interval for a LEAKY BUCKET COUNTER may also be configurable.

It must be possible to change most configuration parameters without a restart. For example, changing a protocol timeout or increasing the size of a pool must not require a restart. However, it is reasonable to require a restart when decreasing the size of a pool; in such cases, a node should simply save the new value so that the operator can defer the reduction until the next restart.

13.1.2 Provisioning

Provisioning refers to craftspeople populating a system with its business objects. Many of these are subscriber profiles, which customize the system's behavior for each of its subscribers by defining, for example, the services that a subscriber is allowed to use.

Provisioning occurs in the administration node, which then downloads subsets of the provisioned data to various nodes through the control node. Provisioned data resides in WRITE-PROTECTED MEMORY when possible. It is also backed up on disks that reside on the control node, so that the control node can reload service and access nodes with their business objects even when the administration node is out of service or otherwise inaccessible.

Because business objects essentially correspond to future work, some form of load balancing must determine, for example, which node will host a particular subscriber. However, load balancing must consider performance and software and hardware restrictions. For example, if a group of business objects will collaborate extensively, assigning them to different nodes could seriously affect capacity. In other cases, the software might not even support their distribution because doing so would be too complex. Similarly, a subscriber who uses a service that requires a specialized hardware device may have to be assigned to one of the few nodes where such devices exist.

13.2 OPERATIONS

This section discusses data that the system (control, service, and access nodes) uploads to administration nodes, so that craftspeople can observe the system's behavior.

13.2.1 Logs

Logs record events that are of interest to people who manage a system. Some logs highlight faults that require manual intervention. Others provide information about the system's behavior for analytical or debugging purposes. The system's documentation must describe each type of log that it generates.

Logs provide information about various system components:

- *Software logs* help designers to debug software faults. In this section, we will focus on logs that are of interest to craftspeople, postponing the discussion of software logs to Section 15.1.
- *Service and resource logs* provide usage statistics for the system's services and resources. The term *operational measurements* describes these usage statistics. Operational measurements are covered in Section 13.2.3 because they are seldom of immediate interest to craftspeople. Although they can verify the system's health, their main purpose is to help determine how many resources the system requires during times of peak load.
- *Hardware logs* highlight faults and state changes in devices such as cards and interfaces. These logs are often of immediate interest to craftspeople. When a hardware failure requires manual intervention, a log identifies the device in question, the reason for its failure, and the corrective action that is required.

Each log provides a severity level so that craftspeople can quickly determine its relative importance:

- *Error logs* highlight situations that require corrective action. A hardware error log, for example, could prompt a craftsperson to replace a faulty device. Software faults also generate error logs, but craftspeople can only resolve them by forwarding them to you and waiting for bug fixes.
- *Warning logs* highlight situations that may require corrective action. They often indicate that the system has detected a problem that may lead to a service degradation. For example, a hardware warning log could flag a transient fault whose recurrence would eventually result in an error log due to the system taking the device out of service. A resource warning log might indicate that all resources in some pool were currently in use, which could signify that the resource was under-engineered. A software warning log identifies an unexpected but nonfatal condition that a software designer should investigate.
- *Informational logs* record events that do not require corrective action. When a craftsperson replaces a device, for example, the system should generate a hardware informational log when it places the new device in service. If the system autonomously corrects a fault, it should generate an information log to highlight this action. Most service and resource logs are informational because their usage statistics relate to normal operation.
- *Debug logs* are only of use to designers. They are usually enabled in the lab. But in the field, they are only enabled while a designer is analyzing a problem.

Designers must classify logs carefully so that craftspeople can focus on logs that require corrective action. They must also ensure that the system will not generate a flood of logs. This degrades capacity and risks overwhelming craftspeople with too much information. It can also lead to the loss of important logs if log buffers overflow. The following guidelines reduce the risk of such outcomes:

- Avoid repeating a log. The exception to this occurs when a log requires corrective action. Such a log should be repeated at an interval determined by its severity and the expected time required to correct the problem. A sequence of logs that report state changes (from state A to B to C to D, for example) should be merged into a single log.
- Avoid generating dependent logs. For example, if a card that hosts a number of interfaces goes out of service, a single log should

be generated. This log might state that the outage has affected *n* interfaces, but a separate log should not be generated for each of those interfaces.

Each processor provides a log framework that applications access through a singleton. The system's overall log framework needs to support the following capabilities:

- *Capturing*. Each log must identify its type (software, service or resource, or hardware) and severity level (error, warning, informational, or debug). For correlation purposes, a log must also include a timestamp and the location where it was generated. It is also helpful to include a sequence number for similar logs.
- *Buffering*. The log framework initially places logs in a buffer in a binary format. The buffer generally operates in a circular manner, but it is undesirable to overwrite error and warning logs if the buffer overflows. Another guideline is to preserve the earlier occurrences of repeated logs so that it will be possible to analyze the original onset of a problem.
- *Spooling and routing*. Access and service nodes send logs to the system's control node. The control node then forwards these logs, as well as its own, to one or more administration nodes for presentation to the craftspeople. The control node should also write all logs to disk to ensure that they will be available for subsequent analysis. However, logs consume a lot of disk space and must therefore be periodically deleted. If craftspeople forget to do this, the log framework must erase old log files to make room for new ones.
- *Suppressing and throttling*. If the system is generating a flood of logs, craftspeople should be able to reduce the volume. A flood is often attributable to a specific type of log. The log framework should therefore support, at a granular level, both suppression (discarding the log) and throttling (reporting only every *n*th occurrence of the log).
- *Formatting*. The conversion of logs from binary to text format should be deferred for as long as possible. Formatting requires CPU time, and formatted logs require more disk space and link bandwidth because they are larger than unformatted logs. Ideally, administration nodes should format logs, although it may be necessary to format them in control nodes. Access and service nodes, however, should not spend time formatting logs.

  Formatted logs should use a consistent terminology so that they are easy to understand. Not only must they be easy to read, they must be easy to parse, because operators usually develop their own log analysis software. If your product is destined for international markets, it should also be able to format logs in various languages

- *Correlating, filtering, and browsing*. The log framework in the administration node should provide tools that make logs easier to analyze.

13.2.2 Alarms

The system generates an ALARM when it enters a state that requires intervention from a craftsperson. Each alarm has a severity level:

- *Critical*. This level usually highlights a widespread loss of service caused by the failure of an essential component. It would also be used to highlight a situation, such as a disk overflow, that prevents the system from generating service usage records that the operator requires to bill subscribers.
- *Major*. This level usually highlights a partial loss of service or a situation, such as the failure of one processor under HOT STANDBY or WARM STANDBY, which would escalate to a critical alarm if another failure occurred.
- *Minor*. This level usually highlights a minor loss of service or a situation which would escalate to a major alarm if another failure occurred.
- *Off*. This level indicates that the alarm has been cleared.

Alarm levels help craftspeople prioritize work when more than one alarm is in effect. An alarm remains in effect until it is cleared, which occurs when a craftsperson takes corrective action to restore the system to normal operation.

The log framework can generate alarms by associating an alarm level with each log. A log must be generated when an alarm is set, escalated, or cleared. A log that clears an alarm has a severity level of *off*. A log that is associated with a critical or major alarm is usually an error log, and a log associated with a minor alarm is usually a warning log.

Extreme systems provide audible and visual indications when an alarm is set. A critical alarm, for example, might cause a bell to ring and a light to flash. If the craftspeople cannot clear the alarm, the system must allow them to suppress the indicators so that they do not become an annoying distraction.

13.2.3 Operational Measurements

OPERATIONAL MEASUREMENTS provide system performance statistics. They allow operators to monitor a system's throughput, calculate how much work it can handle, and determine how many resources are required to handle times of peak load. The term *operational measurements* is commonly used in the telecom domain. Other domains often use the term *performance monitoring* instead.

Application software generates operational measurements to provide the following types of information about services or resources:

- *Event counts*. A telephone switch, for example, reports the total number of call attempts, as well as how many calls were abandoned before answer, how many resulted in busy tone, how many were actually answered, and how many failed because a resource was unavailable or suffered a fault. For hardware devices, and software resources such as object data blocks, operational measurements provide information about how many times the system tried to allocate a resource and how many of these attempts failed because the resource's pool was exhausted.
- *High- and low-water marks*. These could include the maximum number of simultaneous sessions, low-water marks for resource pools, and high-water marks for the length of work queues.
- *Histograms*. These types of operational measurements are obtained by samples taken at regular intervals, perhaps as often as every 10 seconds or every minute. They can include information such as the number of resources in each state (out of service, idle, or in use) and the current length and delay (latency) for each work queue.

The operational measurement framework reports statistics in informational logs that it generates at regular intervals. CONFIGURATION PARAMETERS tell the framework how often to generate each log. Their values are typically in the range of 5 to 15 minutes.

A pair of registers typically implements each operational measurement. The first register gathers data as the system runs. When it is time to report the data, the operational measurement framework copies the first register to the second one and resets the first register. In this way, data can be logged and spooled in the background, during the time that the system gathers data for the next interval. All registers must be copied and reset at the same time to avoid skewing the results reported by different operational measurements.

For a more detailed discussion of operational measurements, see [DEL98].

13.3 MAINTENANCE

MAINTENANCE software monitors the system's health and intervenes when faults occur. It primarily deals with hardware components, although it partially views a failure caused by faulty software as a hardware failure in the affected card.

Maintenance actions can be initiated both autonomously, by the system itself, and manually, by craftspeople. Maintenance software therefore faces tension between MINIMIZE HUMAN INTERVENTION and PEOPLE KNOW BEST [ADAM96]. Initiating maintenance actions autonomously allows the system to react to faults more quickly and makes it easier to operate. On the other hand, maintenance must allow craftspeople to override autonomous actions when they believe that human intervention would expedite fault recovery.

13.3.1 Handling Faults

Fault handling by maintenance software involves a number of steps:

- *Detection.* The maintenance system must quickly detect and react to faults. Software that generates a hardware error or warning log must also inform the maintenance system of the problem. Maintenance then runs a diagnostic test to determine if the component is faulty. This usually involves taking the component out of service, but it is also desirable to provide diagnostics that can run while a component is in service. When in-service diagnostics exist, maintenance should run them at regular intervals to uncover faults proactively.
- *Isolation.* Maintenance must take a faulty component out of service so that it will not affect the rest of the system. Sometimes it is difficult to identify exactly which component is faulty. The maintenance object model must capture hardware dependencies in a directed graph. The graph provides a fault hierarchy that supports fault correlation. If, for example, a hardware device subtends other devices, faults in more than one subtended device may point to a fault in the higher-level device. Fault correlation affects hardware design because a complex design can make it difficult to identify which component is faulty.
- *Notification.* When maintenance removes a component from service or places it in service, it must notify interested observers of this action. For craftspeople, it must generate a log; it may also need to set or clear an alarm. If the component is a device in a resource pool, maintenance must inform the pool manager so that it can remove the device from the idle queue or send a failure message to the application that is using the device. If the component is a node, maintenance must inform other nodes so that they will stop or start sending work to the affected node.

- *Correction.* When maintenance detects a hardware fault, it runs a diagnostic to determine if the component is truly faulty. If the test passes, maintenance returns the device to service. If the test fails, it removes the component from service and attempts to reset it to correct the problem. If maintenance cannot correct the problem, it generates a log to prompt craftspeople to replace the device. Maintenance must therefore identify the faulty device to the level of a single field-replaceable unit. If it cannot do so, it must generate a list of candidates, but this is undesirable because craftspeople will typically replace all of them to correct the fault as quickly as possible.
- *Recovery.* When a component fails and a standby is available, maintenance must divert work to the backup component. This may involve initiating a failover.

13.3.2 High-Level Design

All nodes contain maintenance software. The control node has overall control of the system. It communicates with service and access nodes to monitor their status and send them maintenance commands. Software in the control node may generate these commands autonomously, but craftspeople can also enter them from the administration node, which then forwards them to the control node for subsequent routing to other nodes. The administration node provides maintenance craftspeople with a view of the system's status. At the highest level, it typically provides a control panel that summarizes information about out-of-service components, active alarms, throughput, and overload conditions.

Maintenance software must always be available because it is the only way, short of a total hardware reset, to recover from faults. Maintenance software therefore runs in its own scheduler faction and allocates objects from pools that are not shared by other applications. The separate faction and pools ensure that other applications cannot interfere with maintenance work by consuming too many resources.

Maintenance software primarily consists of state machines that communicate with messages. Each state machine represents a maintainable entity. Many of them correspond to replaceable components such as cards or links. Others implement logical components such as nodes, where a node consists of two cards that run under a HOT STANDBY or WARM STANDBY arrangement.

Two state machines support each maintainable entity. The master runs in the control node, and the slave runs in the node that actually contains the underlying hardware. The terms 'master' and 'slave' are somewhat misleading because the slaves should run as autonomously as possible. This offloads work from the control node and may allow the system to survive an outage in the control node. Commands from the master, however, usually take priority over autonomous actions undertaken by the slave. The master–slave arrangement is an example of HIERARCHICAL CONTROL [DOUG03].

Because the control node runs in active-standby mode, it actually has two copies of each master state machine. If a slave also runs in a node that uses an active-standby mode, there will also be two copies of each slave. Keeping these copies synchronized is important so that the correct state of each maintainable entity is available in the event of a failover.

Device states must be sufficiently granular to support maintenance software. The initial design of some systems, for example, only defines an *out of service* state, which cannot be exited without replacing the device. When the system later adds diagnostics, this causes a problem because a device can now be temporarily out of service for diagnosis. Moreover, the diagnostic may have been initiated by the system itself or by a craftsperson. These states must be further distinguished so that faults identified by the diagnostic can be routed to the maintenance software or craftspeople. Routing these faults to the usual fault handling software could cause an infinite loop in which the maintenance system repeatedly tries to remove the device from service to run diagnostics.

13.3.3 Commands

Maintenance commands equate to signals in a protocol. The subset of signals that a state machine supports depends on its role (master or slave) and the type of component that it implements. Typical commands include

- *Query*, to obtain a component's current state. The overall state is either *in service* or *out of service*, but these states are often qualified. An in-service component may only be partially in service because of faults in subtended components. A component may be out of service as the result of system (autonomous) or manual (craftsperson) action, or because out-of-service links rendered it unreachable.
- *Disable*, to take a component out of service if it is faulty or if it needs to be subjected to invasive procedures that would affect its users.
- *Load*, to download software, firmware, or patches. This command must prevent components from being loaded with incompatible

versions of software that would make it impossible for them to communicate, for example.

- *Test*, to run a diagnostic. An extreme system should adapt diagnostics used during manufacturing for use in live systems. It is difficult to pinpoint and confirm faults in a system that lacks diagnostics. Consequently, craftspeople will often replace many cards to correct a fault, causing a more widespread service disruption. They then ship all of these cards to you because they have no way to tell which ones are faulty. When you test them, most of them result in 'no fault found', so you ship them back. This is a colossal waste of time for everyone involved, so plan to include diagnostics.
- *Abort*, to stop an in-progress *Load* or *Test* command. These commands take time to complete, so it must be possible to interrupt them to perform other work.
- *Reset*, to initiate a software restart or place hardware in an initial state. Each component must unconditionally support this command.
- *Failover*, to force a failover in an active-standby configuration.
- *Enable*, to place a component in service. The system should never place a faulty component in service.

All commands send acknowledgments so that any pending command can be issued as soon as the component finishes handling the first command. Most commands also generate logs to record their use.

Maintenance software must be able to prioritize and preempt commands:

- Prioritization is critical when many faults occur in a short time. For example, if one component subtends another, a fault in the higher-level component may cause a fault to be reported in the lower-level component. Even if the fault in the lower-level component is not spurious, resolving the fault in the higher-level component is more important.
- Preemption prevents in-progress commands from delaying work that has become more important. For example, a *Disable* command always preempts an *Enable* command. Commands issued by craftspeople usually preempt those autonomously issued by the system. After a craftsperson issues a *Disable* command, the system no longer performs autonomous actions on the affected component.

To support prioritization, maintenance defines message priorities that are more granular than those described for overload controls in Section 10.1. To support preemption, maintenance state machines perform an automatic *Abort* when higher priority work arrives.

Commands like *Disable*, *Reset*, and *Failover* often disrupt service for some subscribers. Craftspeople must be warned of these disruptions and prompted to confirm these commands. Other invasive commands, such as some forms of *Load* or *Test*, should be rejected unless preceded by *Disable*.

13.3.4 Hardware Requirements

Maintenance requirements affect hardware design. We already mentioned the need for diagnostics: out-of-service diagnostics to confirm faults, and in-service diagnostics to identify faults proactively. This section discusses some additional requirements.

Maintenance software must be notified when a craftsperson inserts or removes a card so that it can immediately react in an appropriate way. The hardware platform must support this capability by generating an interrupt when a card is inserted or removed. It should also provide a way of identifing the type of card that resides in a slot. Slots and cards must be part of the maintenance object model. Certain slots, for example, may be reserved for specific types of cards, and a card's type may restrict its role (as a control, service, or access node).

Hardware interrupts must be used sparingly so that they do not flood a processor with work. For example, a link interface card should not generate an interrupt on each error. Instead, it should increment a counter that the software can read at its convenience.

Maintenance software faces its greatest challenge when it must recover from a total system outage. A loss of power is the worst possible scenario, which is why batteries and generators usually provide backup power to extreme systems. When a card is inserted or regains power, it must broadcast a request so that the control node will download its software. After a total outage, reloading all cards takes a considerable amount of time. It is therefore desirable to support multicast loading so that cards receiving the same software can be loaded in parallel.

13.3.5 Detecting Outages

The maintenance system must detect, or be informed of, outages as quickly as possible so that it can immediately initiate recovery procedures. We have already mentioned that the hardware platform should generate an interrupt when a card is removed. In other cases, a LEAKY BUCKET COUNTER signals a failure when a hardware device suffers too many faults within a given interval.

For software components, HEARTBEATING often provides outage detection. The approach is simple: a component must regularly send an 'I'm alive' message to the entity that maintains it. For example, a slave maintenance object for a node might send heartbeat messages to its corresponding master in the control node. Heartbeating is a pure software implementation of WATCHDOG. In systems that do not support or use `SIGCHLD` signals, a thread may also send heartbeats to its parent thread.

Heartbeating can incur too much overhead in large systems. Although it takes little time to actually process a heartbeat message, it incurs the full overhead of a message at I/O level. An extreme system typically wants to detect outages within 1 or 2 seconds, so a system containing dozens of cards will burden its control node with dozens of heartbeat messages per second. The same problem can arise if threads send heartbeats to their parents.

One way to avoid heartbeating between processors is to include the messaging system in outage detection. The messaging system in the control node simply sets a flag when it receives a message from a specific node. Once per second, it clears the flag. If the flag was not set as a result of normal work flowing through the system, the messaging system informs the maintenance system that the node may be out of service, at which point the maintenance system can prompt it for a heartbeat. This approach eliminates most heartbeat messages.

Heartbeating between threads can be also be implemented with flags. A parent sets a flag for a child thread, which must then clear the flag every second.

When the slave maintenance object for a node decides to initiate a restart or failover, it should immediately inform the control node of this fact rather than relying on heartbeating. This reduces the interval in which other nodes continue to send messages to the failed node. Recall that the control node informs all the other nodes when a node fails. The messaging systems in the other nodes then nack any messages destined for the failed node.

When a node is under maintenance control (for example, when it is being brought into, or taken out of, service), the messaging system should suppress all non-maintenance messages to that node. It is therefore insufficient for the messaging system to view a node as simply being *available* or *unavailable*. It must also support a *restricted* state. Entry to, and exit from, this state is controlled by maintenance software. If a non-maintenance application tries to send a message to a node in the *restricted* state, the messaging system immediately nacks it, just as if the node were out of service.

13.4 COMMERCIALLY AVAILABLE HARDWARE AND MIDDLEWARE

During the late 1990s, a number of hardware and middleware products began to emerge to speed the development of extreme systems.

Sending configuration parameters, provisioned data, and logs between various nodes is essentially a data distribution problem that will lead you to develop a distributed database. For example, if a node requires data but is out of service, the data must be buffered and sent to it later. Products such as Polyhedra, Solid, and Times Ten provide distributed databases that are sufficiently robust and efficient to be considered for use in extreme systems. These products can serve as an underpinning of your OAM&P system.

General-purpose computing shelves are also available. In many cases, they also provide interprocessor messaging systems and specialized cards for applications such as telecommunications. Firms that offer such platforms include Continuous Computing, Force Systems, and Motorola Computer Group. In some cases, these platforms also offer middleware for developing shelf maintenance software. This type of middleware is also available from GoAhead, a pure software firm.

A number of firms also offer contracting services for developing software that runs in administration nodes. This specialization has arisen because firms that develop extreme embedded systems often lack expertise in developing user interfaces. Because operators primarily interact with an extreme system through its administration nodes, a well-designed graphical user interface will be a strong selling point.

13.5 SUMMARY

- OAM&P software allows craftspeople to populate a system with data and monitor its operation. It also handles hardware faults, either autonomously or in response to commands from craftspeople.
- CONFIGURATION PARAMETERS allow craftspeople to engineer system resources and to control various aspects of a system's behavior.
- Provisioning software allows craftspeople to populate the system with its business objects, such as subscriber profiles.

- The administration node sends CONFIGURATION PARAMETERS and provisioned objects to the control node, which forwards them to the appropriate service and access nodes. This data usually resides in WRITE-PROTECTED MEMORY.
- LOGS notify craftspeople of important events within the system. They include software logs, service and resource logs, and hardware logs.
- Each log is classified as an error, warning, informational, or debug log.
- The log framework captures and buffers logs. It then routes them to the administration node, where it formats them and writes them to disk. It also provides craftspeople with log suppression, throttling, correlation, filtering, and browsing capabilities.
- ALARMS notify craftspeople of situations that require corrective action.
- Each alarm has a severity level: critical, major, minor, or off (when resolved).
- The log framework generates alarms by allowing each log to include an alarm severity level.
- OPERATIONAL MEASUREMENTS are statistics that allow craftspeople to monitor a system's behavior, determine its throughput, and engineer its resources for times of peak usage.
- Operational measurements include event counts, high- and low-water marks, and histograms. They are reported in informational logs.
- Service and access nodes send logs, alarms, and operational measurements to the control node, which forwards them to the administration node. The control node also generates some of this information itself.
- MAINTENANCE software monitors the system's health and reacts to faults, either autonomously or in response to craftsperson commands.
- Maintenance supports fault detection, isolation, notification, correction, and recovery.
- The control node coordinates maintenance activities for the system. Each service and access node also contains maintenance software that manages local components and interacts with the central maintenance software in the control node.
- Maintenance allows craftspeople to query, disable, load, test, reset, and enable components. It also allows them to force failovers and abort a command that is in progress.
- Maintenance software must always be available. It therefore runs in its own scheduler faction and uses its own OBJECT POOLS.

- Maintenance software must be able to prioritize and preempt work.
- Maintenance must detect node outages quickly. It may use HEARTBEATING for this purpose, but this introduces processing overheads. A more efficient approach is to have the messaging system notify maintenance of any node that has not sent a message within a specified interval.

14

Software Optionality

In some situations, software must be conditionally included in, or excluded from, a system. The need for software optionality arises for several reasons:

- *Product families*. Building extreme software is a significant undertaking. It takes time for it to achieve its availability, reliability, capacity, scalability, and productivity goals. Consequently, when the software in one product is on its way to meeting these goals, a firm often decides to reuse it in new products. Many of the frameworks will be reusable, even if the applications are not. Over time, these frameworks end up supporting a product family. However, each product contains different applications and may not need all of the frameworks. This is especially true in small systems, which often have limited memory. Hence the need for software optionality.
- *Targeting*. Within a product family, different products may select different compilers, operating systems, and hardware platforms. Software shared by these products must then be targeted. An abstraction layer typically supports targeting. It defines interfaces that each product implements differently, to account for differences in low-level components. Each software load only needs one implementation of the abstraction layer, so it omits the others.
- *Feature optionality*. A product often ships the same software to all customers because this is much easier than building a custom load for each of them. Consequently, some customers receive capabilities for which they have not paid. These capabilities must be disabled to prevent them from being used for free. Unlike the previous examples, this form of optionality occurs at run time rather than at compile time. Run-time optionality simplifies the build process

Robust Communications Software G. Utas
 ISBN: 0-470-85434-0 (HB)

and allows a capability to be enabled once the customer pays for it. A new software release does not have to be installed.
- *Debugging tools*. Extreme systems include debug tools in production loads to support debugging in the field. These tools are disabled by default and can only be enabled by customer support engineers.

This chapter discusses ways to implement software optionality, both at compile time and at run time.

14.1 CONDITIONAL COMPILATION

The use of `#ifdef` and other CONDITIONAL COMPILATION directives is undoubtedly the most popular technique for implementing software optionality in C and C++ systems. Unfortunately, many systems grossly overuse this technique, often to the point where they become incomprehensible, unmaintainable, and untestable [LAK96]. Wherever possible, conditional compilation should be avoided in favor of one of the other techniques discussed in this chapter.

There is one situation, however, where conditional compilation is appropriate, namely, when targeting software to a platform in which memory is at such a premium that the software simply will not fit unless some of it is removed.

It may seem puzzling that a system could actually contain extra software that can be omitted in this way. However, this situation can arise when deploying software in a small, embedded system, such as a mobile phone. Although a mobile phone does not satisfy all the criteria for an extreme system, it does satisfy some of them. Consequently, its software might be developed using classes that are also used in larger systems. Furthermore, small, embedded systems are often developed offline, on platforms that provide more memory and better debugging capabilities. When the software is finally deployed, however, many of these debugging capabilities are removed. They no longer serve any purpose because problems in these types of systems are debugged in the lab, not in the field. Conditional compilation is then a reasonable way of removing the custom debugging software that is only needed in lab versions of the software.

As another example, many classes in a large system may use an OBJECT POOL. However, because a system with object pools often uses more memory than one that only uses a heap, a small system may want to avoid them. Conditional compilation can then remove the overrides of operators `new` and `delete` that support object pools.

14.2 SOFTWARE TARGETING

Enhancing makefiles to include or exclude software is sometimes the best way to implement SOFTWARE TARGETING. A good example of this is targeting software to different operating systems. The best way to do this is to define a set of wrapper classes that, collectively, comprise an abstraction layer. Although all targets share the class interfaces, their implementations differ from one operating system to the next, so they reside in separate directories. The file organization looks something like this:

```
rtos/Rtos.h
rtos/linux/Rtos.cpp
rtos/ose/Rtos.cpp
rtos/qnx/Rtos.cpp
rtos/vxworks/Rtos.cpp
rtos/win/Rtos.cpp
```

A compile-time symbol then specifies the target operating system. The makefile in the `rtos` directory uses this symbol to select the subdirectory that provides the `Rtos.cpp` file. This approach is far cleaner than liberally populating the software with things like `#ifdef __linux`. However, it may require some refactoring to ensure that most of what appears in the `Rtos.cpp` files is truly platform specific.

You should define an abstraction layer and implement software targeting even if you only plan to deliver your product on one platform. This will allow you to test software offline (on PCs, for example). Using an abstraction layer right from the outset is easier than trying to define one later and then having to retrofit all your software to use it.

Using separate makefiles for each member of a product family allows them to exclude unused interfaces and implementations from their builds. This approach is somewhat cleaner than adding an `#ifdef` to each optional file. Nonetheless, both approaches theoretically require designers to compile each product to ensure that optional capabilities remain optional. It is all too easy inadvertently to create a dependency from a mandatory component to an optional one, thereby breaking the compile of a product that omits the component that was intended to be optional. Ideally, the structure of software directories should prevent such mistakes, but this may not always be possible. However, once a firm is using software across an entire product line, it probably has the resources to develop tools that prevent these mistakes by enforcing dependency restrictions.

14.3 RUN-TIME FLAGS

Although debug capabilities are sometimes excluded from small systems, this is rarely appropriate in large systems that need to support debugging in the field. However, any production load must disable debug tools by default. It can do so with RUN-TIME FLAGS that default to `false`. Setting a flag then enables a tool, allowing it to be temporarily used in the field:

```
if(debugOn)
{
   // debug code
}
```

Flags also surround defensive software, such as extensive sanity checks, whose capacity impact makes it desirable to disable them in the field. These flags allow the defensive code to be reenabled if unforeseen problems arise and the code is needed to help debug problems or preserve system sanity.

Flags can also surround optional capabilities. If the customer does not want, or has not paid, for a capability, a flag disables it. Later, if the customer pays for the capability, it can be enabled by entering a password to flip the flag.

14.4 SUMMARY

- Software needs to be optional when not all members of a product family use it, when it supports a specific platform, when it provides an optional feature, or when it implements debugging capabilities.
- CONDITIONAL COMPILATION is best used to exclude portions of software not required by all members of a product family.
- SOFTWARE TARGETING modifies makefiles to select platform-specific software or to exclude entire source files from products that do not require them.
- RUN-TIME FLAGS are the best way to disable optional features and debugging capabilities.

15

Debugging in the Field

There's a problem in the field. The system is handling hundreds of transactions each second. How are you going to debug the problem? With a breakpoint debugger that halts the system? That'll make an impression! Or maybe print everything that happens to the console? This might allow a *few* transactions per second instead of none, but it will nevertheless create an immediate overload condition and a widespread denial of service.

Now what? You could wait for a subscriber to call the network operator, complaining that some scenario doesn't work. After a while, someone may figure out how to reproduce the problem so that you can debug it in the lab, but this doesn't always happen. Even when it does, it often takes a while to identify the scenario, because subscribers can't always explain precisely who was doing what when the problem occurred.

In many development teams, breakpointing and writing to the console are standard practice. We have already ruled these out for field use, but even in the lab, they have limitations. Breakpointing can cause message timeouts. Writing to the console is fine during unit testing, although it wastes time on recompilation. However, software submitted to the library should *never* write to the console. Not only is it rude to fill the console with garbage, but writing to the console means that there is a bug. And why is a bug being checked in to the code base?

Debugging in the lab is difficult and tedious enough, even when printing to the console and breakpoint debugging are available. The sheer volume of code and the number of messages flowing through the system make it hard to pinpoint problems quickly. A call between

Robust Communications Software G. Utas

two mobile phones, for example, might contain 40 messages and 6000 function calls, so you need many debugging weapons in your arsenal.

This chapter describes techniques that support debugging in live systems. However, these techniques must also be useful in the lab, because they will be the *only* techniques available in the field. If they are inadequate for debugging most lab problems, they will also be inadequate for use in the field.

Debugging facilities in live systems must be thoroughly tested to ensure that they do not harm the system. They must be highly reliable, and they must not slow the system down significantly. They fall into one of two categories: logs and trace tools.

15.1 SOFTWARE LOGS

As mentioned in Section 13.2.1, logs usually highlight events that could require operator intervention. However, software logs are rarely meaningful to craftspeople. They cannot take corrective action to resolve them – except by forwarding them to you and waiting for bug fixes.

Extreme systems seldom use core dumps for debugging. For one thing, SAFETY NET prevents most crashes. Second, when a node gets into trouble, it typically reinitializes itself, using ESCALATING RESTARTS to return to service quickly. Under this strategy, there is neither time to take a core dump, nor does a crash occur. Small executables might use core dumps, but most executables in an extreme system are large. Their core dumps take a long time to generate and analyze, and are therefore only used as a last resort, if other techniques fail to provide sufficient debugging information.

An extreme system's software logs must therefore be comprehensive enough to allow designers to pinpoint and resolve most problems. Truly difficult problems may also require the use of trace tools during debugging. However, customers are often reluctant to allow debugging in live systems. In many cases, they only grant access when a high-profile problem has gone unresolved for some time.

Anything that the software writes to the console (or, more generally, to a log file) prompts questions from the operator. Because a live system generates a considerable number of logs, it must eliminate unnecessary logs. For software, this means disallowing informational logs. They clutter the console and waste time that should be spent on payload work. As a result, the system should only produce three types of software logs: errors, warnings, and object dumps.

15.1.1 Software Error Log

A SOFTWARE ERROR LOG provides information about a trap – a signal or an exception. Although signals usually occur unexpectedly, software throws exceptions deliberately when it detects a condition from which it cannot recover. For example, application software might notice that an object is unexpectedly missing or in an unexpected state. It should then be able to call a static function that simply throws a general application exception. This function should allow the application to provide some data that may be useful during post-mortem analysis. Generally, it is sufficient to provide a reason and an offset. The reason could be the identifier of an unexpected state or event. The offset is simply a sequence number within the function. When a function detects different types of errors, the offset helps to identify the location where it spotted the error:

```
void Thread::SoftwareError(uint reason, uint offset)
{
   throw ApplicationException(reason, offset);
}
```

As discussed in Section 8.1, `Thread::Start` provides a SAFETY NET to catch signals and exceptions that would otherwise be fatal. It subsequently invokes a virtual `Thread::Recover` function to clean up the work in progress. Before it invokes `Recover`, however, it generates a software error log that captures the following information:

- The name of the system and processor in which the error occurred and the time when it occurred. This information must appear in all logs for correlation purposes.
- The thread that was running when the exception occurred.
- The exception itself.
- The thread's stack at the time of the fatal error. The stack is disassembled to provide a **stack trace**: the chain of function calls that led to the exception, along with each function's stack (arguments and local variables).

Unfortunately, satisfying this last requirement is something of a black art. In the same way that C++ does not mandate the capability to throw an exception from a signal handler, it does not mandate the capability to capture and disassemble a thread's stack. Java provides the latter capability, but C++ systems must implement it in a compiler-specific way. Nonetheless, an extreme system must provide

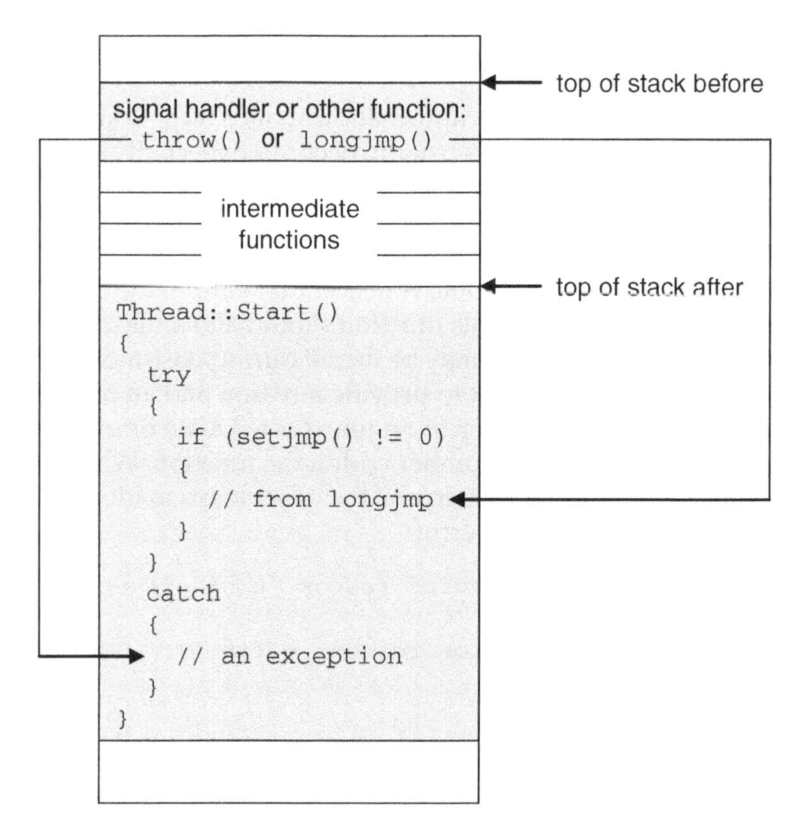

Figure 15.1 Capturing a stack trace for debugging purposes. When a signal handler or other function throws an exception or invokes `longjmp`, the stack is still intact. The stack must be captured at that point because it unwinds before it reaches the safety net in `Thread::Start`.

stack traces in order to offer much hope of quickly resolving bugs. In many cases, a stack trace allows a knowledgeable designer to quickly determine the scenario that led to the problem.

Capturing a stack trace can be difficult (see Figure 15.1). When `Thread::Start` catches an exception, it's too late: the stack has already been unwound. That is, the C++ exception handling mechanism has returned through all of the functions on the stack. Yet the stack is still there, so why not capture it anyway? Unfortunately, this probably won't work. First, calling a function to capture the stack tramples the bottom portion of it, making it difficult or impossible to follow the chain of functions that resides above. Second, when the exception handling mechanism unwinds the stack, it invokes destructors for any local objects in each function. Invoking these destructors also tramples the stack that existed at the time when the exception was thrown.

To overcome these problems, we must capture the stack as soon as a signal or exception occurs. The signal handler must therefore capture the stack *before* it invokes `longjmp` (see Section 8.1.2). When throwing an exception, the exception's constructor can capture the stack, which was the motivation for defining the abstract class `ExceptionWithStack` in Section 8.1.5. Note, however, that when a third-party library, such as the Standard Template Library (STL), throws an exception, the stack will *not* be captured. It would be better if such libraries raised signals so that the signal handler could capture the stack. If you plan to use such libraries, try to change them to capture the stack.

15.1.2 Software Warning Log

A SOFTWARE WARNING LOG provides information about a nonfatal error. It highlights a situation in which software believes that it may not produce the correct result, but one that is not severe enough to abort the work being performed.

A static function, similar to the one used to throw an exception, provides a facility for generating a software warning log:

```
void Thread::SoftwareWarning(uint reason, uint offset)
{
   capture the thread's stack;
   generate a log with the function call chain,
      reason, and offset;
}
```

The warning log is an abbreviated form of the SOFTWARE ERROR LOG. It omits an exception, because there isn't one, and its stack trace simply provides the chain of function calls, omitting each function's arguments and local variables to save time and reduce the size of the log. Nonetheless, this abbreviated stack trace usually helps to determine the scenario that led to the problem.

Many systems define assert macros for generating software error and warning logs. Capitalized macros that evoke C and scream out from the source code offend me, but I am atypical in this regard. If you use assert macros, define them so that they capture reasons and offsets, which are invaluable for diagnosing problems.

Because warning logs do not abort work in progress, designers must use them judiciously to prevent a single problem from causing a flood of logs. For example, a function that receives an out-of-bounds array index as an argument might return `false` or NULL. If it also generates a warning log, callers of the function should usually avoid generating a redundant log.

An AUDIT should generate a log when it recovers an orphaned resource. When the audit recovers idle resources after fixing a corrupt free queue, it should avoid generating a flood of logs by providing a single log that summarizes how many resources it recovered.

15.1.3 Object Dump

When `Thread::Start` handles a fatal error, it first collects the information needed to produce a software error log. It then invokes the `Recover` function so that the thread that trapped can clean up any resources associated with the work that it was performing.

For an invoker thread, recovery means cleaning up the state machine that was handling the failed transaction. Before it cleans up the objects associated with the state machine, however, the invoker thread should first log them in an OBJECT DUMP that provides additional debugging information. The invoker thread maintains a data member that references the top-level object in the running state machine's object model, but now it must find all of the other objects involved in the transaction. It does so by invoking `GetSubtended` (see Section 4.1) on the top-level object. Now that the invoker thread has a list of all the objects involved in the transaction, it generates a log for each one by invoking its `Display` function. This is far more efficient than generating a core dump. It also focuses on the objects that are likely to have caused the problem and provides a symbolic display of each one.

15.2 FLIGHT RECORDER

If a restart clears a node's log buffer, important debugging information is lost. It is therefore desirable to ensure that important logs survive restarts. To this end, each node should provide a FLIGHT RECORDER that serves the same purpose as the 'black box' on an aircraft. If something goes wrong, the flight recorder can later be retrieved for a post-mortem analysis.

Although a flight recorder can be implemented by writing logs to disk, a restart might occur before the log framework manages to spool all buffered logs. It is therefore better to implement the recorder in a memory segment that survives restarts. The flight recorder can use a circular buffer that is of sufficient size to hold, say, 30 minutes of important logs. The buffer can reside in RAM that is not reinitialized during restarts. Another option is to place it in flash memory, if writing to flash is not too CPU intensive.

The flight recorder should capture the following information:

- Error and warning logs, particularly those that set alarms. Serious software problems (traps, restarts, and failovers) must also be included.
- Informational logs that clear alarms or indicate that an error or warning log was successfully resolved.
- Service-affecting actions performed by craftspeople, such as taking a device out of service or installing a software patch.

15.3 TRACE TOOLS

Trace tools are far more invasive than logs. Consequently, they are disabled by default. When enabled, they capture data while the software runs, saving this data in a buffer. A background thread then displays or analyzes the data.

There is so much activity in a live system that trace tools must support triggers. A trigger specifies criteria that, when satisfied, temporarily start or stop a tool so that it only captures a specific transaction or all transactions that occur in a specific session. In this way, the tool only traces scenarios of interest. If the tool were simply turned on globally, its overhead would cause an excessive capacity degradation, and the volume of work performed by the system would quickly overflow the tool's trace buffer, making it easy to lose data associated with the scenario under investigation.

For example, the trace tool framework could define triggers that start or stop a selected set of tools when

- a specific thread (or type of thread, in the case of a THREAD POOL) is running;
- a specific function is invoked;
- a specific message (defined by its protocol and signal) arrives;
- a message arrives from a specific IP address or on a specific IP port;
- a specific type of state machine runs, or enters a specific state or handles a specific event.

15.3.1 Function Tracer

A FUNCTION TRACER captures function calls. This tool is so useful that it quickly becomes the primary means for developers to debug difficult problems in the lab. In the field, however, it must be used carefully because it slows the system down appreciably when enabled.

The function tracer requires a very large trace buffer that contains a record of each function call. There are two ways to capture function calls:

1. By calling a trace function at the top of each function that is to be included in the trace.
2. By using a trace exception, if the CPU provides one. Whenever the CPU executes a subroutine call, subroutine return, or jump instruction, it generates a hardware exception. The function tracer provides an exception handler that records the program counter and the type of instruction (call, jump, or return). After tracing ends, the tool maps each program counter to a function name and offset.

The benefit of the first approach is that, perhaps surprisingly, it typically incurs less processing overhead. Its drawback is that each function must actually invoke the trace function to be captured. However, this is also beneficial because uninteresting functions can avoid capture. By contrast, the second approach is highly invasive. Whereas the first approach slows a system down by a factor of two or three, the second one usually slows it down by a factor of at least ten. It does, however, capture everything.

The first approach is supported by a simple function that can be inlined:

```
void Tracer::ft(const string *func)
{
   if(FunctionTracerOn)
   {
      Thread *thread = Thread::RunningThread();
      if(thread->TraceOn())
         new FunctionTraceRecord (func, thread);
   }
}
```

A function adds itself to the trace as follows:

```
const string ClassNameFunctionName =
                "ClassName::FunctionName";

<return-type> ClassName::FunctionName(<argument-list>)
{
   Tracer::ft(&ClassNameFunctionName);
   // code follows
}
```

Each `FunctionTraceRecord` records the following information:

- A pointer to the function's name (the `func` argument).
- The running thread (the `thread` argument).
- The current time in ticks. When the trace begins, the tool records the time of day and the current time in ticks. This later allows it to convert each function call's tick time to a time-of-day value.
- The function call depth, obtained by counting the number of function call stack frames that reside between `FunctionTraceRecord`'s constructor and the bottom of the stack – yet another example of platform-specific, black-art software.

After tracing stops, this information allows the function tracer to output the sequence of function calls in a format that looks like this:

```
START OF TRACE on SystemName 15-Dec-2003 22:10:30.000

mm:ss.nnn Thr Total Net Function
--------- --- ----- --- --------
10:30.000   6  3712  32 InvokerThread::Enter
10:30.000   6    24  16   InvokerThread::FindWork
10:30.000   6     8   8     Message::Exqueue
10:30.000   6  3640  72   InvokerThread::ProcessWork
10:30.000   6  3568 120     AnApplicationClass::
                                ProcessWork
...
10:30.003   6     8   8   Thread::MsecsLeft
10:30.003   6    40  16   Thread::Pause
10:30.003   6    24  16     Thread::
                                EnterBlockingOperation
10:30.003   6     8   8       Thread::Unlock
END OF TRACE
```

As with logs, the trace banner indicates when and where it occurred. The times in the left-hand column and the banner line are to the millisecond, whereas function times are in microseconds. In this example, a tick is 8 microseconds, so all function times are multiples of this value. Each line shows the running thread and the total and net times spent in the function. Function names are indented to show who called whom. This greatly improves the trace's readability, which is why it is important to capture each function's depth on the stack.

A function tracer with these capabilities can be implemented in about a week. Its cost is recouped quickly, and it eventually pays for itself thousands of times over. It saves a staggering amount of debugging and recompilation time when developers start to abandon breakpointing and `cerr`. It also provides a detailed view of the system's behavior, which is useful both for training purposes and, as we will see later, for improving capacity.

Although it is beyond the scope of this book to provide source code for a function tracer, the following points should help you to implement one:

- `FunctionTraceRecord` derives from an abstract `TraceRecord` that allows all trace tools to be integrated into a common framework. In the next section, we will discuss a message tracer. When function and message tracing are used together, their records can be placed in a common buffer if they subclass from `TraceRecord`. Its constructor records the tick time so that it is included in all records.
- `FunctionTraceRecord` overrides operator `new` to allocate space in the trace buffer. The buffer operates in a circular fashion so that it captures the most recent activity if the buffer wraps around.
- Each `FunctionTraceRecord` contains a pointer to a function name (4 bytes), a thread identifier (2 bytes), a tick time (4 bytes), and a depth (1 byte). It also contains a `vptr` (4 bytes), the overhead for any object that inherits virtual functions. When padded, its size is therefore at least 16 bytes. A trace buffer capable of capturing 8000 function calls therefore needs to be 128 kB in size. For field use, much larger buffer sizes must be supported.
- The total time for a function at depth n is the difference between that function's tick time and the tick time of the next function at depth n.
- The net time for a function at depth n is its total time minus the total time of all functions at depth $n+1$, up to the next function at depth n.
- If a function invoked while constructing a `FunctionTraceRecord` tries to include itself in the trace, a stack overflow occurs!

The following types of commands control the function tracer:

- `set buffsize` *n* allocates a trace buffer of *n* kB.
- `include thread` *n* includes function calls in a specific thread.
- `include thread all` selects function calls in all threads.
- `exclude thread` *n* excludes function calls in a specific thread.
- `startat func` *f* starts tracing when a specific function is invoked.

- `stopat func` *f* stops tracing when a specific function is invoked.
- `query selections` lists the selected threads and functions.
- `start` initiates tracing.
- `stop` ends tracing.
- `query buffer` shows how much of the buffer is occupied by entries.
- `show trace` *f* writes the trace to file *f*.
- `clear buffer` erases all entries in the buffer.
- `clear selections` deselects all threads and functions.

The `startat` and `stopat` commands are primarily for field use. They help to capture specific scenarios, which reduces both the tool's overhead and the risk of buffer overflows.

15.3.2 Message Tracer

A MESSAGE TRACER captures messages that are received or sent by software under observation. It captures the entire contents of each message, whether the message is external (entering or leaving the system) or internal.

The message tracer is implemented by adding capture logic to functions such as `Message::Send` and `InvokerThread::ReceiveMsg`. If a `Buffer` object holds each physical message, the easiest way to capture a message is to clone this object. A `MessageTraceRecord`, derived from `TraceRecord`, contains the tick time (from the `TraceRecord`) and a pointer to the cloned `Buffer` object.

The commands that control function tracing must now be enhanced to support message tracing:

- `set tool` *s* `on` selects the tool(s) identified by string *s*.
- `set tool` *s* `off` deselects the tool(s) identified by string *s*.
- `query tools` lists the tools that have been selected.
- `include addr` *a* includes messages from/to a specific IP address.
- `include port` *p* includes messages from/to a specific IP port.
- `startat msg` *p s* starts tracing when a message matches a specific protocol and signal.
- `stopat msg` *p s* stops tracing when a message matches a specific protocol and signal.

For example, `set tool fm on` selects both function and message tracing.

Here are some points to consider when implementing a message tracer:

- The message tracer interleaves its records with function trace records when both tools are enabled. Because the message tracer clones `Buffer` objects, its records must be deleted to prevent memory leaks. Deletion occurs when the trace buffer wraps around and when clearing the trace buffer.
- Messages encoded in text format are easy to read. Messages encoded in binary, however, simply appear in hex by default. Consequently, implementing functions to display binary messages in text form saves a lot of debugging time. This is straightforward if each message header identifies the message's protocol: the buffer can then be passed to a `DisplayMsg` function that the protocol provides. For a TLV MESSAGE, `DisplayMsg` displays the message header and invokes a `DisplayParm` function on each parameter.

15.3.3 Tracepoint Debugger

Although a FUNCTION TRACER and MESSAGE TRACER are adequate for debugging most field problems, truly difficult problems still require a debugger. In the field, however, you need a TRACEPOINT DEBUGGER that uses tracepoints rather than breakpoints. When a user defines a tracepoint, the debugger also asks for trace instructions, which resemble opcodes. When the debugger reaches a tracepoint, it interprets the instructions to capture memory contents in a buffer. For example, one instruction might save part of the stack, and another might save an object by dereferencing a pointer. After the debugger captures the requested data, it resumes execution of the system's software rather than halting.

15.4 SUMMARY

- An extreme system needs field-safe debug tools. This rules out breakpoint debugging and writing to the console.
- A SOFTWARE ERROR LOG captures the entire stack when a signal or exception occurs.
- A SOFTWARE WARNING LOG captures the function call chain when a software component detects a nonfatal error.
- An OBJECT DUMP logs the contents of objects that were cleaned up as the result of a signal or exception.

- A FLIGHT RECORDER supports the post-mortem analysis of outages by saving important logs and events in a memory segment that survives restarts.
- Detailed debugging requires trace tools. To be useable in the field, these tools must implement triggers that enable them in specific scenarios.
- A FUNCTION TRACER captures the sequence of function calls at run-time.
- A MESSAGE TRACER captures external and internal messages at run-time.
- A TRACEPOINT DEBUGGER interprets instructions to capture specific data structures at run time. It immediately resumes execution of the software after each tracepoint.

16

Managing Capacity

Developers of an extreme system must be able to predict how much throughput it will provide. Operators require this information in order to assess a system's cost effectiveness. They therefore ask for it in advance, both when considering which system to purchase and when writing a contract. Furthermore, they often seek guarantees as to the system's future capacity so that they can plan the growth of their networks. As a system adds capabilities, it is natural for its capacity to erode slowly. If this erosion is excessive, however, operators need to purchase more systems than they anticipated. Thus, they may insist that a new software release not decrease capacity by more than 2 or 3%. Consequently, you must regularly assess a system's capacity to ensure that it will continue to meet its obligations. The earlier you detect an excessive degradation, the earlier you can mount a recovery effort.

Most of this chapter describes tools that help to predict capacity and how to use them once a system has been developed. However, there is something else to consider before you deploy a system for the first time, so we will begin our discussion there.

16.1 SET THE CAPACITY BENCHMARK

The initial release of a product establishes a permanent capacity benchmark. As long as the system's initial capacity makes it cost effective in comparison with competing products, it will probably be acceptable. However, even a 5% degradation in a future software release will foster customer complaints. It is therefore prudent to *not* optimize the initial release more than necessary, and even to make

Robust Communications Software G. Utas
 ISBN: 0-470-85434-0 (HB)

it inefficient in known ways. Future releases can then remove these inefficiencies and undertake further optimizations. Recovering lost capacity often involves considerable effort, so it is a good idea to give yourself a cushion right at the outset. This strategy allows you to add new capabilities without significantly degrading capacity.

16.2 MEASURING AND PREDICTING CAPACITY

To manage capacity, you need tools that measure processing costs. These tools identify which areas would benefit the most from optimization. They also help to predict the maximum workload that the system can support. This section describes three profiling tools that serve these purposes.

16.2.1 Transaction Profiler

The cost of payload work is the primary determinant of a system's capacity. A TRANSACTION PROFILER measures the cost of this work. It instruments payload software to obtain performance measurements, which are then fed into a model of the system's workload to predict its capacity. The model accounts for the types of transaction that occur in customer sites and the relative frequencies of those transactions. The initial version of the model is based either on customer predictions or on the behavior of similar systems that are already in service. Later, OPERATIONAL MEASUREMENTS obtained from the system when it is in service update the model (see Section 13.2.3).

If an unacceptable capacity degradation is discovered late in the development cycle, the software release will be late. It takes time to recover lost capacity, after which the system must be retested. Using the transaction profiler early and often during the development cycle prevents this stressful situation. It allows you to predict capacity in advance so that you can take corrective action immediately. Initially, you can predict capacity by feeding the transaction profiler's timing information into a model of the system's workload. Later during the development cycle, you must verify these predictions by running the system under load to observe its actual throughput.

Two components determine the cost of payload work:

1. *I/O cost*: the cost of receiving a message and placing it on a work queue.
2. *Transaction cost*: the cost of processing the message.

A transaction profiler must measure these costs with a high degree of precision. It does so by taking timestamps when work begins and ends. It is implemented as a trace tool that places records in the same trace buffer used by the FUNCTION TRACER and MESSAGE TRACER. The transaction profiler creates an I/O record as soon as an I/O thread receives a message, and it creates a transaction record as soon as an invoker thread takes a message off a work queue. These records contain the following data:

- The tick time when the record was created.
- The protocol and signal associated with the message.
- A pointer to the object (such as a state machine) that received the message.

The transaction profiler saves a pointer to the last record that it created. This allows it to reaccess the record just before the I/O thread waits for its next message, or just before the invoker thread dequeues its next message. At that time, the transaction profiler obtains the current tick count, which is the time at which the work ended, and saves it in the record that it created at the beginning of the transaction. In that record, the difference between the initial and final tick counts yields the cost of the work. For even greater accuracy, the transaction profiler can modify one of the tick counts to remove its own overhead.

In many cases, payload work consists of a series of transactions that implement some service. Saving the message's protocol and signal, and the object that received it, allows these transactions to be correlated and added together to determine the total time required to provide the service.

As a concrete example, we will look at predicting the call handling capacity of a basic telephone switch. A switch supports two basic types of interfaces: lines (subscriber telephones) and trunks (links for routing calls between switches). The switch therefore supports the four basic call types shown in Figure 16.1:

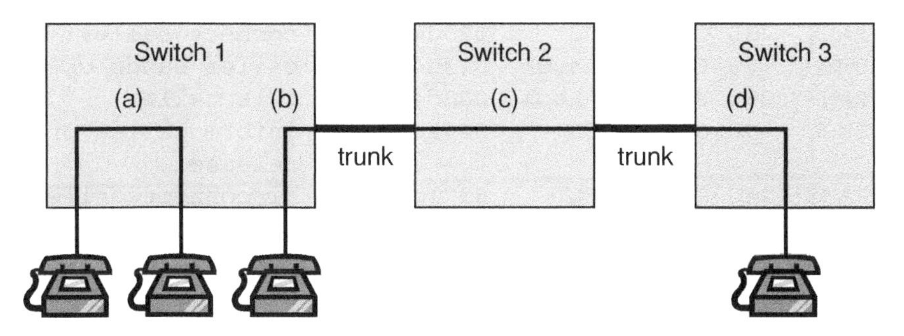

Figure 16.1 Basic call types in telephone switches. (a) line-to-line; (b) line-to-trunk; (c) trunk-to-trunk; (d) trunk-to-line.

1. Line-to-line: a direct call between two subscribers served by this switch.
2. Line-to-trunk: a call from a subscriber served by this switch to a subscriber served by another switch.
3. Trunk-to-line: a call from a subscriber served by another switch to a subscriber served by this switch.
4. Trunk-to-trunk: a call between two subscribers served by other switches, but routed through this switch.

Let's say that the relative frequencies of these call types is

Line-to-line	50 %
Line-to-trunk	20 %
Trunk-to-line	20 %
Trunk-to-trunk	10 %

For a line-to-line call, the transaction profiler might provide the following data:

```
Event Time Proto Signal  Object
----- ---- ----- ------  ------
IOMSG  170 POTS  Offhook 023d4f60 // initiate call
TRANS  610 POTS  Offhook 023d4f60 // provide dial tone
IOMSG   70 POTS  Digits  023d4f60 // address arrives
TRANS 1660 POTS  Digits  023d4f60 // determine callee
IOMSG  170 CIP   IAM     0271c7e0 // receive call
TRANS  850 CIP   IAM     0271c7e0 // ring callee
IOMSG   70 CIP   EOS     023d4f60 // acknowledge call
                                  // setup
TRANS  190 CIP   EOS     023d4f60 // ringback to
                                  // caller
IOMSG   70 POTS  Offhook 0271c7e0 // callee answers
TRANS  230 POTS  Offhook 0271c7e0 // connect callee
IOMSG   70 CIP   ANM     023d4f60 // inform caller of
                                  // answer
TRANS  230 CIP   ANM     023d4f60 // connect caller
IOMSG   70 POTS  Onhook  023d4f60 // caller hangs up
TRANS  890 POTS  Onhook  023d4f60 // idle caller
IOMSG   70 CIP   REL     0271c7e0 // inform callee of
                                  // release
TRANS  340 CIP   REL     0271c7e0 // disconnect callee
IOMSG   70 POTS  Onhook  0271c7e0 // callee hangs up
TRANS  560 POTS  Onhook  0271c7e0 // idle callee
```

The call contains two processing contexts: the caller's call half, and the callee's call half. Each call half receives POTS (plain ordinary telephone service) messages from one subscriber and exchanges internal

CIP (call interworking protocol) messages with the other call half to coordinate control of the call. Each IOMSG records the cost of receiving a message and placing it on a work queue, and each TRANS records the cost of processing that message. The total cost of the call is 6770 microseconds, or 6.77 milliseconds. After we run the transaction profiler on the other call types, we might end up with the following results:

Call Type	Frequency	Cost (msec)
Line-to-line	50 %	6.77
Line-to-trunk	20 %	6.31
Trunk-to-line	20 %	5.85
Trunk-to-trunk	10 %	4.46

The cost of an average call is therefore

$$(0.5 \times 6.77) + (0.2 \times 6.31) + (0.2 \times 5.85) + (0.1 \times 4.46) = 6.263\,\text{msec}.$$

If the processor dedicates 80% of its time to handling calls, then under this particular model it can process

$$(0.8 \times 3{,}600{,}000\,\text{msec/h})/(6.263\,\text{msec/call}) = 459{,}843\,\text{calls/h}$$

Thus, if the system needs to support 1.5 million calls per hour, it requires four processors.

Although a transaction profiler's main purpose is to provide inputs to capacity models, it also has another purpose. If you save its results in a database and compare them over time, you can determine which transactions slowed down. Then, by comparing old and new function traces of these transactions, you can determine exactly which functions slowed down. This comparison helps you to identify software that you need to speed up in order to recover lost capacity.

16.2.2 Thread Profiler

A THREAD PROFILER collects statistics about each thread. How much CPU time did the thread use? How many times was it scheduled in? What was the average length of time for which it ran? Most operating systems provide such a profiler, so it is unlikely that you will need to develop one yourself. Doing so is difficult in any case, because it typically requires hooks in the scheduler. You could implement it for locked threads, however, by adding hooks to `Thread::Lock` and `Unlock`.

A thread profiler is useful for assessing the performance of nonpayload threads. Because these threads only receive a modest amount of CPU time, they seldom require the detailed analysis provided by a TRANSACTION PROFILER. Nonetheless, you should periodically assess their performance to ensure that the system can still dedicate the same high percentage of its time to payload work.

16.2.3 Function Profiler

Sometimes the FUNCTION TRACER helps you to identify functions that have slowed down significantly and which you can speed up to recover lost capacity. It is common, however, to discover that many functions have slowed down by small amounts. The question is then how to speed up the system. Here, the 80/20 rule applies. That is, the system probably spends 80% of its time in 20% of the functions. It could even be that it spends 50% of the time in 5% of the functions. Focusing on these functions is the fastest way to improve capacity.

A FUNCTION PROFILER identifies high-cost functions. It does so by post-processing the output of the FUNCTION TRACER to generate the following information for each function:

- How many times the function was called.
- The total net time spent in the function.
- The average net time spent in the function.
- The minimum net time spent in the function.
- The maximum net time spent in the function.

This information helps you to determine which functions offer the greatest potential for improving capacity. Saving t microseconds in a function that is called n times is usually easier than saving nt microseconds in a function that is called once. The most frequently used functions are usually very short. Making them inlines is one way to improve capacity quickly.

16.3 SUMMARY

- An extreme system must manage its capacity because its customers do not want to buy more systems in order to offset reductions in capacity.
- Your first software release establishes a capacity benchmark, so don't make it any faster than it needs to be.

- A TRANSACTION PROFILER measures the CPU time used by each unit of work at the I/O and application levels. It is the primary tool for accurately measuring the costs of payload work. The costs are fed into a model of the system's workload to predict its capacity, which must be verified by running the system at peak load.
- A THREAD PROFILER measures the CPU time used by each thread. It is most often used to monitor the cost of nonpayload threads.
- A FUNCTION PROFILER post-processes the output of a FUNCTION TRACER to provide data about each function's cost and how often it is called. This data focuses a capacity improvement program on speeding up software that is executed most frequently or that has slowed down significantly.

17

Staging Carrier-Grade Software

Now that we have looked at techniques for implementing carrier-grade software, we need to discuss how to introduce them into your system.

The first part of this chapter offers some guidelines for selecting third-party software. You need to choose this software carefully because it could make it difficult or even impossible to attain your reliability, availability, or capacity goals, or to implement some of the techniques described in this book.

The last part of this chapter discusses which of techniques in this book you need to implement in your first release and which ones you can defer to a subsequent release.

17.1 SELECTING AN INFRASTRUCTURE

At an early stage in your initial software release, you need to select the third-party software on which you will build your system. This activity can proceed in parallel with software development to the extent that you can develop and test software on PCs, for example. However, after a short time, it is prudent to start running your software on its target platform, integrating it with the third-party software that you have chosen.

Robust Communications Software G. Utas
 ISBN: 0-470-85434-0 (HB)

17.1.1 General Guidelines

Third-party software includes compilers, operating systems, libraries, and the types of middleware discussed in Section 13.4. Here are some guidelines for choosing such software:

- Prefer software for which source code is available. This will allow you to determine how it works or even to debug it. Eventually, you may need to modify it to improve its reliability or capacity. If you don't have access to the source code, ensure that the third-party software supplier will be responsive to your requests for assistance.
- Ask if the software creates its own processes or threads. If it does, obtain a list of them, along with their priorities. This will help you to integrate the software when using Cooperative Scheduling or Proportional Scheduling.
- Find out which functions can block. If you need to use these functions often, particularly during payload work, see if nonblocking versions of them exist. A lot of third-party software assumes that scheduling you out is no big deal, but it is when you're using Cooperative Scheduling to run to completion. Such software can put you on the slippery slope to ubiquitous semaphores.

  For example, an in-memory database typically blocks when it accesses a locked record. Some applications, however, would prefer to have the lookup fail rather than block, particularly when the probability of accessing a locked record is low. The database should support this with an argument that tells the lookup function to return `false` rather than block.

17.1.2 Compiler

If you have a choice of compilers, use your payload software to benchmark them. Compilers can have a significant effect on capacity because some are much better at optimizing code than others. Beyond that, prefer a compiler that helps you implement the following techniques, some of which may also require support from the operating system:

Technique	Requirement
Safety Net	Throwing exceptions from signal handlers
Safety Net	Cleanup of objects during `longjmp` (if signal handlers cannot throw exceptions)
Hitless Patching	Incremental loader for replacing a function's object code

17.1.3 Operating System

Choosing an operating system will probably be your most difficult decision. I am unaware of any commercially available operating system that satisfies all of the requirements in the following list. To the extent that you can find one that does, you will reduce your development costs.

Technique	Requirement
COOPERATIVE SCHEDULING	Thread locking (to run unpreemptably)
COOPERATIVE SCHEDULING	RUN-TO-COMPLETION TIMEOUT
COOPERATIVE SCHEDULING	SIGVTALRM (if RUN-TO-COMPLETION TIMEOUT is not built in)
PROPORTIONAL SCHEDULING	Thread factions
PROPORTIONAL SCHEDULING	Ability to change timewheel at run-time
PROPORTIONAL SCHEDULING	Faction inheritance (to resolve faction inversion)
STACK OVERFLOW PROTECTION	Signal when thread overruns stack
LOW-ORDER PAGE PROTECTION	Signal when dereferencing NULL pointer
USER SPACES	Multiple threads per process
WRITE-PROTECTED MEMORY	Supported by memory management calls
SAFETY NET	Alternate stack for signal handler
RELIABLE DELIVERY	Fast reporting of connection loss and buffer overflow
FLIGHT RECORDER	Memory segment that survives reboots
FUNCTION PROFILER	Microsecond granularity for raw clock time

17.1.4 Debugging

A few sections in this book mentioned the need for debugging capabilities whose implementation is something of a black art. The more of these that you can buy, the fewer you will need to implement.

Technique	Requirement
SOFTWARE ERROR LOG	Disassembled full stack trace
SOFTWARE WARNING LOG	Disassembled partial stack trace (functions only)

Function Tracer	Library call that returns the depth of the current function
Tracepoint Debugger	Support for tracepoints, not just breakpoints

17.2 ADDING CARRIER-GRADE TECHNIQUES

Some of the techniques described in this book belong in your first software release, but you can defer others to subsequent releases. Your initial release needs to include any technique that meets either of the following criteria:

1. It affects the programming model. Postponing its use would be undesirable because it would cause many changes in application software when added.
2. It significantly improves availability, reliability, capacity, or scalability. Using it in your initial release is therefore important for achieving a reasonable level of software quality.

If a technique does not meet one of the above criteria, you can implement it later. Nevertheless, you may need to design your software to anticipate its eventual inclusion. Other techniques, however, can be added transparently, with little or no preparation or changes to applications.

17.2.1 Techniques for Release 1.0

There should be a good reason for omitting any of the following techniques from your first release. Some of them are fundamental to a carrier-grade programming model, so trying to add them later would cause a lot of rework. Others are critical for operating a system in the field, and others significantly improve productivity.

Technique	**Purpose**
`Object` Class	Defines basic functions for most objects
Singleton	Improves capacity by creating objects during system initialization
Flyweight	Improves capacity by sharing an object
Polymorphic Factory	Delegates object creation

Thread CLASS	Provides THREAD-SPECIFIC STORAGE and a WRAPPER FACADE for threads and implements extreme thread techniques
COOPERATIVE SCHEDULING	Avoids critical region bugs and improves capacity
HALF-SYNC/HALF-ASYNC	Separates I/O and applications
DEFENSIVE CODING	Improves reliability and availability
INITIALIZATION FRAMEWORK	Provides structure for main
TLV MESSAGE	Provides message abstraction and improves capacity
PARAMETER TYPING	Avoids bugs when building messages
PARAMETER FENCE	Detects trampling when building messages
CONFIGURATION PARAMETERS	Supports site-specific requirements
LOGS	Provides status information to craftspeople
ALARMS	Prompts craftspeople to take corrective action
OPERATIONAL MEASUREMENTS	Tracks system performance
MAINTENANCE	Recovers from outages and allows craftspeople to initiate corrective action
SOFTWARE WARNING LOG	Speeds up lab and field debugging
OBJECT BROWSER	Speeds up lab and field debugging
FUNCTION TRACER	Speeds up lab and field debugging
MESSAGE TRACER	Speeds up lab and field debugging
SET THE CAPACITY BENCHMARK	Simplifies resolution of future capacity degradations
TRANSACTION PROFILER	Measures processing costs to help predict capacity
SOFTWARE TARGETING	Supports different platforms
RUN-TIME FLAGS	Disables or enables software at run-time

The initial release of a distributed system must also include whichever of the following techniques it plans to use. Some of these are not critical, but trying to introduce them later would cause too much upheaval.

Technique	Purpose
HEARTBEATING	Detects outages
RELIABLE DELIVERY	Simplifies applications
MESSAGE ATTENUATION	Avoids message floods
PREFER PUSH TO PULL	Reduces cost of data accesses
HOMOGENEOUS DISTRIBUTION	Improves scalability
HIERARCHICAL DISTRIBUTION	Separates soft and hard real-time applications
HALF-OBJECT PLUS PROTOCOL	Reduces frequency of interprocessor messages
LOAD SHARING	Improves availability of stateless applications
COLD STANDBY	Improves availability of stateful applications

Some techniques require judicious use. If you need them, it is easier to include them at the outset; adding them later would cause too much rework. In some cases, this is because they require special hardware.

Technique	Purpose
THREAD POOL	Removes blocking operations from invoker threads
USER SPACES	Firewalls applications
SHARED MEMORY	Reduces capacity impact of USER SPACES
CALLBACK	Improves capacity by eliminating messages
HETEROGENEOUS DISTRIBUTION	Provides scalability
SYMMETRIC MULTI-PROCESSING	Provides scalability
HOT STANDBY	Improves availability of stateful applications
MEMORY CHECKPOINTING	Supports hitless failover
VIRTUAL SYNCHRONY	Supports hitless failover

Ideally, your initial release should include the following techniques. However, the conflicting forces of time constraints and the customer's quality expectations will determine how many you implement immediately and how many you defer. All of them can be introduced with little or no change to applications.

Technique	Purpose
OBJECT POOL	Avoids fragmentation, enables an object AUDIT, and speeds up object creation
OBJECT NULLIFICATION	Detects users of deleted objects
RUN-TO-COMPLETION TIMEOUT	Detects infinite loops in locked threads
RUN-TO-COMPLETION CLUSTER	Avoids state-space explosions in state machines
STACK OVERFLOW PROTECTION	Detects thread stack overflows
LOW-ORDER PAGE PROTECTION	Detects users of NULL pointers
LEAKY BUCKET COUNTER	Tracks frequency of faults to trigger corrective action
SAFETY NET	Recovers from errors that would be fatal
AUDIT	Recovers orphaned resources and fixes queue corruptions
PARAMETER TEMPLATE	Avoids bugs when building messages
FINISH WHAT YOU START	Prevents crashes during overload
DISCARD NEW WORK	Prevents crashes during overload
IGNORE BABBLING IDIOTS	Protects against a source that generates too many messages
SOFTWARE ERROR LOG	Speeds up lab and field debugging
OBJECT DUMP	Speeds up lab and field debugging

17.2.2 Techniques for Release 2.0 and Beyond

The following techniques must appear in Release 2.0:

Technique	Purpose
PROTOCOL BACKWARD COMPATIBILITY	Prerequisite for ROLLING UPGRADE
ROLLING UPGRADE	Supports gradual installation of a new release
OBJECT REFORMATTING	Reformats data to its schema in a new release

The following techniques can be introduced in Release 2.0 or later. They either improve availability or debugging.

Technique	Purpose
Proportional Scheduling	Guarantees CPU time to all work
Watchdog	Detects outages
Escalating Restarts	Returns a processor to service more quickly
Binary Database	Speeds up reload restarts and software upgrades
Throttle New Work	Prevents work sources from overloading a server
Warm Standby	Improves availability of stateful applications
Application Checkpointing	Supports hitless failover
Object Checkpointing	Supports hitless failover
Hitless Patching	Installs bug fixes with no service impact
Hitless Upgrade	Installs new releases with minimal service impact
Flight Recorder	Captures critical events to support analysis of outages
Maintenance	Provides diagnostics
Tracepoint Debugger	Speeds up field debugging

The remaining techniques improve capacity and can therefore be introduced as required:

Object Template	Speeds up object creation
Quasi-Singleton	Speeds up object creation
Embedded Object	Speeds up object creation
Object Morphing	Avoids cost and complexity of deep copying
Parameter Dictionary	Reduces cost of parameter searches
In-Place Encapsulation	Reduces message copying
Stack Short-Circuiting	Reduces cost of intraprocessor messages
Message Cascading	Reduces message copying
Message Relaying	Reduces message copying
Eliminating I/O Stages	Reduces cost of receiving messages
No Empty Acks	Eliminates messages
Polygon Protocol	Eliminates messages
Function Profiler	Identifies functions to speed up
Thread Profiler	Identifies threads to speed up

18

Assessing Carrier-Grade Software

This chapter defines maturity levels for carrier-grade software. The purpose of these levels is to provide a framework for assessing software that claims to be, or needs to become, carrier grade. This framework can guide due-diligence and self-assessment exercises.

Due-diligence activities seldom include a meaningful assessment of a product's software architecture or programming model. One reason for this is that people who perform due diligence often do not understand the degree to which software affects carrier-grade attributes. Some of them believe that rigorous testing and a sufficiently mature development process are enough to produce a carrier-grade product. The reality, however, is that the wrong programming model will *never* produce a carrier-grade product. You can test it all you want, continually adding code to fix problems, but you will never attain carrier-grade levels of availability, reliability, capacity, and scalability until you redesign the software.

Even when those who perform due diligence realize the importance of software architecture, they usually lack specific knowledge of the programming models and techniques that produce carrier-grade products. They may want to perform a software assessment, but they're not sure what to look for and what to ask.

If you want to assess a product's software to determine if it is, or can become, carrier grade, the previous chapters in this book should suggest many areas for investigation. This chapter outlines these areas and defines four maturity levels based on the techniques that a system uses in each area:

Robust Communications Software G. Utas

ISBN: 0-470-85434-0 (HB)

- **Level 0**. This system is not carrier grade. To reach Level 1, it may need to change its programming model, which could involve significant rework.
- **Level 1**. This system provides basic carrier-grade capabilities and should be able to evolve to higher levels.
- **Level 2**. This system provides additional carrier-grade capabilities, to the point where it should satisfy most customers.
- **Level 3**. This is a mature carrier grade system that should satisfy even the most demanding customers.

The areas that we will assess are

- *Language*: the programming language in which the software is written.
- *Object management*: techniques for managing objects.
- *Tasks*: the longevity of processes and threads.
- *Critical regions*: how the programming model deals with them.
- *Scheduling*: the system's scheduling disciplines.
- *I/O*: how the system performs I/O.
- *Messaging*: techniques for messaging.
- *Fault containment*: techniques for reducing the impact of software faults.
- *Error recovery*: techniques for recovering from software errors.
- *Overload*: techniques for handling overload situations.
- *Processor failures*: techniques for handling processor failures.
- *Operability*: techniques for improving the system's operability.
- *Software installation*: techniques for installing software.
- *Debugging*: techniques for debugging software in the field.

Because we define maturity levels for each area, it is possible for a system to exhibit different levels of maturity in different areas. Some areas do not specify techniques at each level. Others specify multiple techniques at the same level, which means that a system could partially meet a level's requirements.

This book contains many techniques whose sole purpose is to improve capacity. Our assessment model omits these techniques because they are not critical in themselves, but only serve to address shortfalls in capacity.

We will now look at the maturity levels that a system can exhibit in each of the above areas.

18.1 LANGUAGE

- Level 0: Little or no use of an object-oriented language.
- Level 1: Extensive use of an object-oriented language.

A system can attain carrier gradeness without using an OO language. However, its journey will be more difficult, and its use of a non-OO language suggests an inadequate focus on software architecture.

18.2 OBJECT MANAGEMENT

- Level 0: Extensive use of the heap.
- Level 1: Object Pool and Object Nullification.
- Level 2: Provides an Audit for each Object Pool.

At Level 0, a system risks gradual memory leaks and fragmentation, to the point where object allocation fails and an outage occurs. At Level 1, a system avoids fragmentation but still risks memory leaks. At Level 2, a system is at low risk for suffering an outage caused by these problems.

18.3 TASKS

- Level 0: Often creates short-lived tasks (processes or threads) at run time.
- Level 1: Creates daemons during system initialization.

At Level 0, a system's capacity suffers, and it risks outages caused by zombie tasks that gradually eat up memory. It needs significant rework to reach Level 1.

18.4 CRITICAL REGIONS

- Level 0: Extensive use of preemptive scheduling and semaphores.
- Level 1: Cooperative Scheduling.
- Level 2: Run-to-Completion Timeout.

A system at Level 0 allows the scheduler to preempt tasks indiscriminately. It is at high risk for critical region bugs, which often lead to memory corruptions and outages. This system may require significant rework to reach higher levels.

A system at Level 1 greatly reduces the risk of critical region bugs. However, it will hang if an infinite loop occurs in a locked thread. A system at Level 2 eliminates the latter risk.

18.5 SCHEDULING

- Level 0: Priority scheduling.
- Level 1: Round robin scheduling.
- Level 3: Proportional Scheduling.

At Level 0, a system risks thread starvation during times of heavy load and may therefore fail to perform important work. It should nonetheless be able to evolve to higher levels that only use priority scheduling tactically.

A system at Level 1 uses the techniques in Section 5.7.3 to markedly reduce the risk of thread starvation. A system at Level 3 eliminates this risk.

18.6 I/O

- Level 0: Applications perform I/O.
- Level 1: Half-Sync/Half-Async.

A system at Level 0 cannot prioritize incoming work. Consequently, it can only implement rudimentary overload controls and is therefore at risk of thrashing and crashing when presented with more work than it can handle. It will require some rework to reach Level 1, which separates I/O from applications to provide greater flexibility in prioritizing work.

18.7 MESSAGING

- Level 0: Often uses synchronous messaging.
- Level 1: Reliable Delivery, Message Attenuation, and Parameter Fence. Almost always uses asynchronous messaging.
- Level 2: Parameter Typing and Parameter Template.

At Level 0, the programming model does not include state machines. The system suffers from the problems discussed in Section 5.5.3. Its applications will have to be rewritten to attain higher levels.

At Level 1, state machines are central to the programming model. Applications rarely need to retransmit internal messages and do not cause message floods. The system detects memory corruptions during message construction, so it is unlikely that such errors will occur in the field.

At Level 2, the system uses techniques to reduce the risk of building messages that contain invalid data.

18.8 FAULT CONTAINMENT

- Level 0: Does not use fault containment techniques.
- Level 1: Defensive Coding, Stack Overflow Protection, and Low Order Page Protection.
- Level 3: Write-Protected Memory.

A system at Level 0 risks memory corruption because it fails to protect key memory areas. It also performs erratically because developers often neglect to write software that handles error paths. It can add software to address these shortcomings. However, handling all the error paths will take time. It's not just a question of adding a lot of missing code, but of fundamentally changing the programming culture.

A system at Level 1 contains faults by handling error paths and by protecting memory areas that face the greatest risk of corruption. A system at Level 3 also safeguards its configuration data, and other objects whose data rarely changes, from corruption.

18.9 ERROR RECOVERY

- Level 0: Reboots or reinitializes to recover from errors.
- Level 1: Initialization Framework and Heartbeating.
- Level 2: Safety Net, Audit, and Watchdog.
- Level 3: Escalating Restarts and Binary Database.

A system at Level 0 recovers from fatal software errors slowly, often after generating a large core dump. In a distributed system, other processors may not even be informed of the failure. This effectively leads to hung resources when a failure destroys objects involved in interprocessor collaborations. To reach higher levels, the system will require some rework.

A system at Level 1 foregoes core dumps to return to service faster. It detects failures quickly and reports them to craftspeople and other processors. However, software traps (such as invalid memory accesses or dividing by zero) cause outages.

A system at Level 2 handles traps by cleaning up work (such as a session) that encounters what would be a fatal error at Level 1. If the cleanup fails, the system provides a second line of defense that recovers orphaned resources. It also uses adjunct hardware to detect and force a reinitialization of a hung node.

A system at Level 3 usually performs a partial reinitialization to return a node to service quickly. If this fails to restore the node to health, the scope of the reinitialization increases.

18.10 OVERLOAD

- Level 0: Thrashes or crashes when presented with too much work.
- Level 1: Finish What You Start and Discard New Work.
- Level 2: Ignore Babbling Idiots.
- Level 3: Throttle New Work.

A system at Level 0 runs into trouble by accepting more work than it can handle. At Level 1, a system prioritizes incoming work so that it will finish what it has accepted before it takes on new work. If it receives too much work, it continues to run close to its peak capacity.

A system at Level 2 denies service to a source that is presenting it with too much work. At Level 3, a system proactively backpressures work sources to prevent them from sending it too much work.

18.11 PROCESSOR FAILURES

- Level 0: Does not provide backup processors to take over work.
- Level 1: Load Sharing and Cold Standby.
- Level 3: Warm Standby or Hot Standby.

A system at Level 0 suffers a total outage if an important processor fails. At Level 1, a system assigns critical work to more than one processor. It therefore remains in service if a processor fails, but sessions in that processor are lost. At Level 3, another processor takes over these sessions without dropping them.

18.12 OPERABILITY

- Level 0: Provides rudimentary administrative software.
- Level 1: Configuration Parameters, Logs, Alarms, Operational Measurements, and Maintenance.
- Level 2: Out-of-service diagnostics.
- Level 3: In-service diagnostics.

At Level 0, a system provides the minimum amount of software that will allow craftspeople to operate it. A system at Level 1 is clearly designed for operability. It provides capabilities that allow craftspeople to configure it, engineer it, monitor its behavior, and initiate corrective action. A system at Level 2 broadens these capabilities and adds the ability to test hardware for faults after removing it from service. At Level 3, the system proactively tests hardware for faults.

18.13 SOFTWARE INSTALLATION

- Level 0: Requires a service outage to install software.
- Level 1: Rolling Upgrade.
- Level 2: Hitless Patching.
- Level 3: Hitless Upgrade

Table 18.1

Area	Level 0	Level 1	Level 2	Level 3
Language	Non-OO	Primarily OO		
Object management	Primarily uses a heap	Object Pool, Object Nullification	Audit for each Object Pool	
Tasks	Often created at run time	Most tasks are daemons		
Critical regions	Contains many semaphores	Cooperative Scheduling	Run-to-Completion Timeout	
Scheduling	Priority scheduling	Round-robin scheduling		Proportional Scheduling
I/O	Performed by applications	Half-Sync/Half-Async		
Messaging	Often uses synchronous messaging	Almost always uses asynchronous messaging; Reliable Delivery, Message Attenuation, Parameter Fence	Parameter Typing, Parameter Template	
Fault containment	None	Defensive Coding, Stack Overflow Protection, Low-Order Page Protection		Write-Protected Memory
Error recovery	None	Initialization Framework, Heartbeating	Safety Net, Audit, Watchdog	Escalating Restarts, Binary Database
Overload	Thrashes and crashes	Finish What You Start, Discard New Work	Ignore Babbling Idiots	Throttle New Work
Processor failures	No backup processors	Load Sharing, Cold Standby		Warm Standby, Hot Standby
Operability	Provides a minimal set of capabilities	Configuration Parameters, Logs, Alarms, Operational Measurements, Maintenance	Out-of-service diagnostics	In-service diagnostics
Software installation	Requires an outage	Rolling Upgrade	Hitless Patching	Hitless Upgrade
Debugging	No field-safe debug tools	Software Warning Logs, Object Browser, Function Tracer, Message Tracer	Software Error Logs, Object Dump	Flight Recorder, Tracepoint Debugger

A system at Level 0 must reinitialize a processor to install software. To install a new software release, all of its processors must be upgraded in parallel. At Level 1, a system supports a serial upgrade of its processors. A system at Level 2 can install patches without disrupting service. At Level 3, a system can install a new software release with minimal service disruption.

18.14 DEBUGGING

- Level 0: Cannot be debugged in the field.
- Level 1: SOFTWARE WARNING LOGS, OBJECT BROWSER, FUNCTION TRACER, and MESSAGE TRACER.
- Level 2: SOFTWARE ERROR LOGS and OBJECT DUMP.
- Level 3: FLIGHT RECORDER and TRACEPOINT DEBUGGER.

A system at Level 0 cannot be debugged in the field without a high risk of a major service disruption. Its developers must try to reproduce problems in the lab based solely on core dumps, problem reports, and *ad hoc* information written to the console.

A system at Level 1 provides standard debug logs that help developers to identify field problems. If developers cannot identify the scenario that led to a problem, the system provides field-safe debug tools that gather more detailed information.

At Level 2, a system generates detailed debug logs when it recovers from an error that is fatal in a Level 1 system. At Level 3, the system includes a debugger for analyzing the most difficult problems. It also records critical events to support the analysis of outages.

18.15 SUMMARY

Table 18.1 summarizes the techniques used in each area at each level of maturity.

Glossary

Abstract factory A design pattern for instantiating an object whose class is not directly known to user of the abstract factory. *See* [GHJV95].

Access node A node that provides external interfaces to subscribers or other systems.

Ack Positive acknowledgment. In a request–response message pair, an ack indicates that the request succeeded.

Active processor The processor that is performing payload work in a HOT STANDBY or WARM STANDBY configuration.

Administration node A node that downloads data to other nodes and which provides craftspeople with a view of the system's behavior.

Alarm An indication given to craftspeople when a hardware or software failure or resource shortage has caused a service outage or degradation for a number of subscribers.

Application checkpointing Adding checkpointing software to applications in *ad hoc* manner. The software sends messages from an active processor to its standby to replicate objects on the standby.

Asynchronous messaging A form of messaging in which the sender's software continues to run rather than being blocked while waiting for a response. It requires the use of state machines. The opposite of synchronous messaging.

Robust Communications Software G. Utas
 ISBN: 0-470-85434-0 (HB)

Audit A background thread that looks for, and typically tries to correct, inconsistencies such as orphaned resources or corruptions in data structures.

Auxiliary data block A block of memory that is appended to an object to hold some data when not all of the object's data will fit into the data block provided by its OBJECT POOL.

Availability The proportion of time that a system is in service.

Babbling idiot An interface that is generating an excessive number of messages.

Blocked thread A thread that cannot run because it is waiting for a blocking operation to finish.

Blocking operation An operation that causes a thread to be scheduled out until the operation completes. Examples include sleeping, acquiring a semaphore, sending a synchronous message, reading a socket, or performing disk I/O.

Bug A software fault that leads to an error or failure.

Callback A function that a client provides to a service so that the service can later invoke the function to inform the client of an event that is of interest to the client. It is often implemented by having the client register an object in a REGISTRY owned by the service.

Capacity The amount of work that a single processor can perform.

Carrier A firm that operates a network of extreme systems.

Carrier-grade A system that provides the high levels of availability, reliability, capacity, and scalability required by a carrier.

Checkpointing Maintaining synchronization between two processors (an active and a standby) that run in HOT STANDBY or WARM STANDBY mode. It allows the standby processor to preserve work during a failover.

Cold standby A survivability strategy that sets aside one or more spare processors to take on the work of any in-service processor that fails.

Concrete factory An object that instantiates objects in a specific class that is not visible to users of the concrete factory. *See* [GHJV95].

Connectionless protocol A protocol in which each incoming message can be processed independently of previous messages. It consists of either datagrams (unacknowledged, unsolicited messages) or request–response message pairs, and is therefore restricted to one or two messages per dialog.

Connection-oriented protocol A protocol that defines phases for connection setup, message exchange, and disconnection. A connection setup request creates a session that must maintain state information to process subsequent requests correctly. Unless the connection is rejected, each dialog consists of three messages (connect, connect ack, and disconnect) plus additional service requests between the connect ack and disconnect.

Context switch Switching execution from one task to another. A context switch can occur during a blocking operation or as the result of preemptive scheduling.

Control node A node that monitors the status of other nodes and initiates recovery actions when a node fails.

Cooperative scheduling A scheduling policy in which threads, rather than the scheduler, determine when context switches occur. The threads generally run to completion. The opposite of preemptive scheduling. *See* [PONT01].

Craftspeople The people who administer and operate a system or network.

Critical region A section of software which accesses data that is also accessed by another thread. A semaphore must protect a critical region because a context switch could cause one thread to access the data while another thread is in the process of modifying it.

Cyclic scheduling A scheduling policy in which each thread runs at a regular interval. Often used in hard real-time systems.

Daemon A task which is created during system initialization and that lives forever, or at least until a restart occurs.

Deadlock A situation in which a group of threads remains permanently blocked because of a circular dependency among blocking operations. A deadlock occurs if, for example, two threads send each other synchronous messages at about the same time.

Dialog A sequence of interdependent messages exchanged in a protocol.

Distribution Dividing work among multiple processors in order to increase a system's overall capacity or survivability.

Dynamic data Data that changes frequently enough to disqualify it from residing in WRITE-PROTECTED MEMORY.

Embedded object An object that resides within another object and whose lifetime is therefore bounded by that of the outer object.

Embedded system A system that is dedicated to a specific purpose.

Error A deviation from specifications (incorrect operation) caused by a fault.

Escalating restarts A strategy for returning a node to service as quickly as possible by only partially reinitializing its software. If the reinitialization fails to keep the node in service, the scope of the next reinitialization increases.

Event An input to a state machine, causing the state machine to perform work.

Event handler Software that performs a state machine's work in reaction to an event. It is usually selected based on the state machine's current state.

Executable *See* Process.

Extreme system A system whose design is constrained by the carrier-grade forces of availability, reliability, capacity, and scalability.

Faction A set of threads that share a minimum percentage of CPU time. Under PROPORTIONAL SCHEDULING, factions replace thread priorities.

Failover Switching work from one processor to another under HOT STANDBY, WARM STANDBY, or COLD STANDBY.

Failure An error that causes the loss of work or renders a component incapable of performing work.

Fault The root cause of an error or failure.

Five-nines A measure of availability in which a system is out of service for no more than roughly 5 minutes per year.

Function tracer A trace tool that captures the sequence of function calls executed by software and displays them in an indented format to show nesting.

Gobbler Software that uses an excessive amount of some system resource.

Half-object plus protocol (HOPP) A design pattern that fully or partially replicates an object in other processors to improve their capacity. *See* [MESZ95].

Half-sync/half-async A design pattern that separates I/O (receiving messages) from the work performed in reaction to those messages. One or more I/O threads queue incoming messages on work queues that are serviced by one or more invoker threads. *See* [SCH96a].

Hard-deadline scheduling The need to run a thread close to, but no later than, a specific time. Often a requirement in hard real-time systems.

Hard real-time system A system that *must* perform work within a specified time to perform correctly.

Heartbeating Sending an 'I'm alive' message to the entity that manages a component to prevent it from declaring the component to be out of service.

Heterogeneous distribution Distributing work by assigning different functions to different processors.

Hierarchical distribution Distributing work by assigning lower level, CPU-intensive functions to some processors (such as access nodes), and higher-level, more complex work to other processors (such as service nodes).

Hitless failover A failover under WARM STANDBY or HOT STANDBY. It preserves work in the standby processor to avoid disrupting service.

Hitless patching Installing a software patch without disrupting service.

Hitless upgrade Installing a new software release without disrupting service.

Homogeneous distribution Distributing work by assigning different subscribers to different processors so that all processors perform the same functions.

HOPP *See* HALF-OBJECT PLUS PROTOCOL.

Hot standby A survivability strategy that provides hardware checkpointing between two processors by verifying synchronization after each CPU instruction.

Input handler Application-specific software that an I/O thread invokes to identify the work queue in which a message should be placed. The appropriate input handler is selected based on the protocol of the incoming message.

Invoker thread A thread that takes incoming messages from a queue filled by an I/O thread and invokes application objects to process these messages.

I/O level Software that is responsible for performing I/O, such as `tnettask` and I/O threads.

I/O thread A thread that receives messages and places them on a work queue for later processing by an invoker thread.

IP port registry A REGISTRY that selects an input handler based on the IP port on which a message arrived.

ISR (interrupt service routine) Software that handles a processor interrupt, such as an I/O interrupt or clock interrupt.

Kernel mode A memory protection strategy in which operating system data structures are made inaccessible to threads except during system calls.

Latency The delay between receiving a request and sending a response, which is primarily determined by how long a request waits in a queue before it is processed.

Leaky bucket counter A design pattern that detects when a certain number of events occur within a certain timeframe. It is often used to track faults so that a system will neither overreact to intermittent, expected faults nor spend too much time handling a flurry of faults. *See* [GAM96].

Load sharing A survivability strategy in which two or more processors share work, such that the remaining processors continue in service if any of them fail.

Locked thread A thread that has acquired the run-to-completion lock to run unpreemptably with respect to other threads that also contend for this lock.

Log An event that is brought to the attention of craftspeople. If the event has caused a service degradation or outage, it also raises an alarm.

Maintenance Autonomous or manual procedures that implement fault detection, isolation, notification, correction, and recovery.

Memory checkpointing Sending modified pages from an active processor to its standby to maintain synchronization.

Message An input that belongs to a specific protocol. Its contents consist of a signal and zero or more parameters.

Message tracer A trace tool that captures entire messages.

Morphing *See* OBJECT MORPHING.

n + *m* **sparing** *See* COLD STANDBY.

Nack Negative acknowledgment. In a request–response message pair, a nack indicates that the request failed.

Network thread A thread that implements a messaging protocol, such as an IP stack.

Node A processor in a distributed system. In HOT STANDBY or WARM STANDBY, it often refers to a pair of processors in an active–standby configuration.

NUC (non-upward compatible) Changing an interface in a way that is not transparent to users of the interface. A NUC change to a class forces its users, at a minimum, to be recompiled. A NUC change to a protocol forces all users of the protocol to upgrade their software.

OAM&P Operations, administration, maintenance, and provisioning.

`Object` **class** The class from which all nontrivial objects ultimately derive. It defines functions that all objects should support.

Object checkpointing Providing a framework for APPLICATION CHECKPOINTING.

Object data block A block of memory allocated by an OBJECT POOL for housing an object created at run time.

Object morphing Changing an object's class at run time.

Object nullification Resetting some or all of an object's data (including its `vptr`) to improve the likelihood of detecting stale accesses to the object.

Object pool An object management strategy that pre-allocates object data blocks in a pool to avoid use of the heap at run time. A class that uses an object pool overrides operators `new` and `delete` to obtain blocks from, and return them to, the pool.

Object reformatting Reformatting data to the schema of a new release to prepare for a software upgrade, or converting checkpointed objects serialized in one release to their format in a new release.

Object template Using a block-copy operation to initialize an object quickly from a pre-allocated instance of that object.

Operational measurements Usage statistics presented to craftspeople to allow them to engineer a system and evaluate its throughput.

Operations Generating logs, alarms, and operational measurements for craftspeople to inform them of a system's behavior.

Operator *See* Carrier.

Outage A failure that causes a loss of service for some or all subscribers.

Overload control A strategy that prevents thrashing by prioritizing, discarding, or throttling work when a processor receives more work than it can handle.

Parameter An argument passed in a message.

Patch Software that is installed to fix a bug.

Payload work Work performed by services.

Placement `new` A version of operator `new` that constructs an object at a specific memory address.

`PooledObject` A subclass of `Object` that uses an OBJECT POOL to manage its objects.

Polymorphic factory An ABSTRACT FACTORY that delegates object instantiation to a CONCRETE FACTORY that resides in a REGISTRY.

POSIX (portable operating system interface) A standard that promotes software portability by defining a uniform set of operating system capabilities.

Preemptive scheduling A scheduling policy in which a context switch occurs as soon as a higher priority thread is ready to run or, under round-robin scheduling, when a thread's timeslice expires and another thread of the same priority is ready to run.

Primordial thread The first thread, the one that executes `main`.

Priority inheritance Resolving priority inversion by temporarily raising a thread's priority when it owns a resource that is blocking a higher priority thread.

Priority inversion A situation in which a higher priority thread cannot run because it is blocked by a lower priority thread. For example, the higher priority thread might be waiting for a semaphore currently owned by the lower priority thread, or it might be waiting for the lower priority thread to respond to a synchronous message.

Priority scheduling A scheduling policy in which a context switch selects the highest priority thread as the next one to run.

Process A thread or group of threads that run in a user space.

Programming model A set of techniques followed by all of the software in a system. Whereas a system's architecture primarily addresses its interfaces, its programming model primarily addresses its implementation.

`ProtectedObject` A subclass of `Object` that creates objects in WRITE-PROTECTED MEMORY.

Proportional scheduling A scheduling policy that assigns threads to factions and guarantees each faction a minimum percentage of CPU time. The opposite of priority scheduling.

Protocol A set of signals and parameters that are used to construct messages, along with rules that define the order for sending and receiving signals and the parameters that are mandatory or optional for each signal.

Protocol backward compatibility Evolving a protocol so that a processor running a new software release can communicate with a processor running a previous software release.

Provisioning Configuring a system by populating it with data.

Quasi-singleton A class whose objects usually behave as SINGLETONS but which makes provision for allocating another instance of an object when the current object is still in use.

Race condition A situation where software must handle the near-simultaneous occurrence of two events to avoid a bug. Examples include critical regions, deadlocks, and two processing contexts sending each other requests at about the same time.

Reactive system A system that responds to messages arriving from external sources.

Real-time system *See* Hard real-time system *and* Soft real-time system.

Registry An object that houses a set of polymorphic objects and which uses an identifier to select the appropriate object when delegating work polymorphically.

Reliability The percentage of time that a system operates free of errors.

Remote procedure call (RPC) A function call that sends a message, usually synchronously, in which case its thread blocks until it receives a response.

Restart Reinitializing a processor to recover from a software error.

Robustness The ability to maintain availability and reliability despite the occurrence of hardware and software faults.

Rolling upgrade Installing a new software release in a distributed system one processor at a time.

Round-robin scheduling A scheduling policy in which threads of the same priority receive a timeslice to perform their work, with a context switch occurring at the end of the timeslice if another thread of equal priority is ready to run.

RPC *See* Remote procedure call.

Run to completion A strategy that reduces critical region bugs by running each thread locked to suppress preemptive scheduling. *See* [DEBR95].

Run-to-completion cluster A set of state machines that collaborate by exchanging high-priority messages. This prevents members of the cluster from having to deal with messages that could arrive from outside the cluster during transient states which arise from their collaboration. *See* [UTAS01].

Run-to-completion lock A global semaphore which must be acquired by a thread that wishes to run to completion. It prevents preemptive scheduling among the threads that acquire it.

Run-to-completion timeout A sanity timeout that occurs when a thread runs locked too long.

Safety net Catching all exceptions and signals in a thread's entry function in order to clean up the thread's current work and reenter or recreate the thread.

Scalability The ability to increase a system's overall capacity by adding more processors.

Service Software that implements a capability used by subscribers.

Service node A node that primarily runs payload work (services).

Session A group of objects that implements an individual dialog in a connection-oriented protocol.

Session processing The processing of connection-oriented protocols.

`SharedObject` A subclass of `Object` that creates objects in a global shared memory segment.

Signal 1. A message type in a protocol. 2. An event that is received by a thread, usually to flag a software error that is fatal unless caught by a signal handler.

Signal handler A callback that is invoked when a thread receives a signal.

Singleton An object that is the only instance of its class. *See* [GHJV95].

SMP *See* SYMMETRIC MULTI-PROCESSING.

Soft real-time system A system that must perform work within a specified time to provide an acceptable response time.

Spinlock A loop that obtains a lock by polling. Only useful in SMP systems.

Stack overflow A fatal error that occurs when a thread overruns its stack, usually by nesting function calls too deeply or by declaring too many large local variables.

Stack short-circuiting A technique which improves capacity by bypassing the lower layers of a protocol stack when sending an intraprocessor message.

Stack trace The chain of function calls that appears in a thread's stack, starting with the thread's entry function and ending at the function that is currently running. It may also include each function's arguments and local variables.

Standby processor The processor that can immediately take over if the active processor fails in a HOT STANDBY or WARM STANDBY configuration.

State An object that defines how far a state machine has progressed in handling a transaction or session. It selects the event handler to invoke when the state machine receives an event.

Stateful Software that must use state machines to handle a sequence of messages in a connection-oriented protocol.

Stateless Software which need not use state machines because it can process each incoming message independently of previous messages.

State machine A set of states, events, and event handlers which implement some capability. A state machine is required to support a connection-oriented protocol.

Static data Data which changes sufficiently infrequently for it to reside in WRITE-PROTECTED MEMORY.

Subscriber An end user of a system, often a customer of a carrier.

Subscriber profile The data required to support a subscriber, such a list of the services that the subscriber may use.

Survivability The ability of a system to remain in service when hardware or software failures occur.

Synchronous messaging A form of messaging in which the sender's thread is blocked while waiting for a response. The opposite of asynchronous messaging.

Symmetric multi-processing (SMP) A hardware configuration in which multiple processors share a common memory.

System initialization Running the software that brings a processor to the point where it is ready to handle payload work.

Task A term that refers to a process or a thread when a concept applies to both of them.

Thread A sequence of instructions that executes on its own stack. A thread shares its address space with other threads that run under the same process.

`Thread Class` A WRAPPER FACADE [POSA00] for native threads which provides THREAD-SPECIFIC STORAGE [POSA00] and supports various extreme software techniques.

Thread pool A group of threads which perform the same type of work when it requires the use of blocking operations. Using multiple threads reduces latency because some (or all) of the work is performed concurrently.

Thread starvation Under priority scheduling, a situation in which lower priority threads are prevented from running because higher priority threads are using all the CPU time.

Tick The time between two clock interrupts that cause the scheduler to run. The length of a tick is system specific. A timeslice in a round-robin scheduler consists of a small number of ticks.

Timeslice Under round-robin scheduling, the number of ticks that the scheduler allows a thread to run before preempting it to run another thread of equal priority.

Timewheel A circular buffer whose entries are visited at regular time intervals to perform some function, such as scheduling threads or servicing timer requests.

TLV (type–length–value) **message** A message that encodes each parameter using a parameter identifier (type), length, and contents (value).

Tracepoint debugger A debugger that collects data at specified instructions and then allows the system to resume execution instead of halting.

Trampler Software that corrupts memory by writing through an invalid or stale pointer or an out-of-bounds array index.

Transaction The work performed in response to an incoming message.

Transaction processing The processing of connectionless protocols.

Trap A signal or C++ exception.

Upgrade Installing a new software release as opposed to a patch.

User spaces A memory protection strategy in which each process runs in a memory segment that is inaccessible to other processes.

Virtual synchrony A strategy that implements WARM STANDBY by passing each input to both the active and standby processors and verifying synchronization by comparing their outputs.

`vptr` (virtual function pointer) A hidden data member of each object that defines or inherits virtual functions. It references an array which contains a pointer to the correct instance of each virtual function that is associated with the object's class.

Warm standby A survivability strategy in which software provides checkpointing between an active and a standby processor.

Watchdog A timer whose expiry is treated as a fatal error. *See* [PONT01].

Worker thread A member of a THREAD POOL.

Write-protected memory A memory protection strategy that places critical data in a memory segment that can be read by all threads but that can only be modified after performing an unprotection operation.

References

[ADAM96] M. Adams, J. Coplien, R. Gamoke, R. Hanmer, F. Keeve, and K. Nicodemus, "Fault-Tolerant Telecommunication System Patterns," in [PLPD96], pp. 549–62. Republished in [RIS01].

[BERN97] P. Bernstein and E. Newcomer, *Principles of Transaction Processing*. Morgan Kaufmann, San Francisco, California, 1997.

[CDKM02] F. Cottet, J. Delacroix, C. Kaiser, and Z. Mammeri, *Scheduling in Real-Time Systems*, John Wiley & Sons, Chichester, UK, 2002.

[DEBR95] D. DeBruler, "A Generative Pattern Language for Distributed Computing," in [PLPD95], pp. 69–89. Republished in [RIS01].

[DEBR99] D. DeBruler, remarks during the TelePLoP hot topic at the 1999 ChiliPLoP conference.

[DEL98] D. DeLano, "Telephony Data Handling Pattern Language," presented at the 1998 PLoP conference. *http://hillside.net/plop/plop98/final_submissions/P53.pdf*.

[DIJK68a] E. Dijkstra, "Cooperating Sequential Processes," in F. Genuys, editor, *Programming Languages*, pp. 43–112. Academic Press, New York, 1968.

[DIJK68b] E. Dijkstra, "Goto Considered Harmful," in *Communications of the ACM*, **11**(3), pp. 147–48, 1968.

Robust Communications Software G. Utas
© 2005 John Wiley & Sons, Ltd ISBN: 0-470-85434-0 (HB)

[DOUG99] B. Douglass, *Doing Hard Time*, Addison-Wesley, Reading, Massachusetts, 1999.

[DOUG00] B. Douglass, *Real-Time UML* (second edition), Addison-Wesley, Reading, Massachusetts, 2000.

[DOUG03] B. Douglass, *Real-Time Design Patterns*, Addison-Wesley, Boston, Massachusetts, 2003.

[ECPP99] The Embedded C++ Technical Committee, *The Embedded C++ Specification*. *http://www.caravan.net/ec2plus*.

[GAM96] R. Gamoke, "Leaky Bucket Counters," in [PLDP96], pp. 555–56. Republished in [RIS01].

[GAR02] J. Garland and R. Anthony, *Large-Scale Software Architecture: A Practial Guide Using UML*, John Wiley & Sons, Chichester, UK, 2002.

[GHJV95] E. Gamma, R. Helm, R. Johnson, and J. Vlissides, *Design Patterns*, Addison-Wesley, Reading, Massachusetts, 1995.

[GNULIB] Free Software Foundation, *The GNU C Library*. *http://www.gnu.org/software/libc/manual*.

[GNUNLE] "Non-Local Exits," in [GNULIB], Chapter 23.

[GNUSIG] "Signal Handling," in [GNULIB], Chapter 24.

[GRAY93] J. Gray and A. Reuter, *Transaction Processing: Concepts and Techniques*, Morgan Kaufmann, San Francisco, California, 1993.

[HAN99a] R. Hanmer and G. Stymfal, "An Input and Output Pattern Language: Lessons from Telecommunications," in [PLPD99], pp. 503–36. Republished in [RIS01].

[HAN99b] R. Hanmer and M. Wu, "Traffic Congestion Patterns," presented at the 1999 PLoP conference. *http://hillside.net/plop/plop99/proceedings/hanmer/hanmer629.pdf*.

[HAN00] R. Hanmer, "Real Time and Resource Overload Language," presented at the 2000 PLoP conference. *http://hillside.net/plop/plop2k/proceedings/Hanmer/Hanmer.pdf*.

[HAN02] R. Hanmer, "Operations and Maintenance 1," presented at the 2002 PLoP conference. *http://hillside.net/plop/plop2002/final/Hanmer-OAM-1.pdf*.

[HAN03] R. Hanmer, "Patterns of System Checkpointing," presented at the 2003 PLoP conference. *http://www.cs.wustl.edu/~cdgill/PLoP03/system_checkpointing.pdf.*

[HOH03] L. Hohmann, *Beyond Software Architecture: Creating and Sustaining Winning Solutions*, Addison-Wesley, Boston, Massachusetts, 2003.

[ITU99] International Telecommunication Union, *Recommendation Z.100: Specification and Description Language (SDL)*, Geneva, Switzerland, 1999.

[LAK96] J. Lakos, *Large-Scale C++ Software Design*, Addison-Wesley, Reading, Massachusetts, 1996.

[LIA99] S. Liang, *Java Native Interface: Programmer's Guide and Specification*, Addison-Wesley, Reading, Massachusetts, 1999.

[MESZ95] G. Meszaros, "Pattern: Half-object + Protocol (HOPP)," in [PLPD95], pp. 129–32. Republished in [RIS01].

[MESZ96] G. Meszaros, "A Pattern Language for Improving the Capacity of Reactive Systems," in [PLDP96], pp. 575–91. Republished in [RIS01].

[MEY92] S. Meyers, *Effective C++: 50 Specific Ways to Improve Your Programs and Designs*, Addison-Wesley, Reading, Massachusetts, 1992.

[MEY96] S. Meyers, *More Effective C++: 35 New Ways to Improve Your Programs and Designs*, Addison-Wesley, Reading, Massachusetts, 1996.

[PAR00] J. Pärssinen and M. Turunen, "Patterns for Protocol System Architecture," presented at the 2000 PLoP conference. *http://hillside.net/plop/plop2k/proceedings/Parssinen/Parssinen.pdf.*

[PAR01] J. Pärssinen and M. Turunen, "Pattern Language for Architecture of Protocol Systems," presented at the 2001 EuroPLoP conference. *http://hillside.net/patterns/EuroPLoP2001/papers/parssinen.zip.*

[PAR02] J. Pärssinen and M. Turunen, "Pattern Language for Specification of Communication Protocols," presented at the 2002 EuroPLoP conference. *http://hillside.net/patterns/EuroPLoP2002/papers/Parssinen_Turunen.zip.*

[PLPD95] J. Coplien and D. Schmidt (editors), *Pattern Languages of Program Design*, Addison-Wesley, Reading, Massachusetts, 1995.

[PLPD96] J. Vlissides, J. Coplien, and N. Kerth (editors), *Pattern Languages of Program Design 2*, Addison-Wesley, Reading, Massachusetts, 1996.

[PLPD98] R. Martin, D. Riehle, and F. Buschmann (editors), *Pattern Languages of Program Design 3*, Addison-Wesley, Reading, Massachusetts, 1998.

[PLPD99] N. Harrison, B. Foote, and H. Rohnert (editors), *Pattern Languages of Program Design 4*, Addison-Wesley, Reading, Massachusetts, 1999.

[PONT01] M. Pont, *Patterns for Time-Triggered Embedded Systems*, Addison-Wesley, Reading, Massachusetts, 2001.

[POSA96] F. Buschmann, R. Meunier, H. Rohnert, P. Sommerlad, and M. Stal, *Pattern-Oriented Software Architecture: A System of Patterns*, John Wiley, & Sons, Chichester, UK, 1996.

[POSA00] D. Schmidt, M. Stal, H. Rohnert, and F. Buschmann, *Pattern-Oriented Software Architecture: Patterns for Concurrent and Networked Objects*, John Wiley & Sons, Chichester, UK, 2000.

[POSIX03] *IEEE Standard 1003.1: The Open Group Technical Standard Base Specifications, Issue 6. http://www.unix-systems.org/single_unix_specification.*

[RIS01] L. Rising (editor), *Design Patterns in Communications Software*, Cambridge University Press, Cambridge, UK, 2001.

[RJB98] J. Rumbaugh, I. Jacobson, and G. Booch, *The Unified Modeling Language Reference Manual*, Addison-Wesley, Reading, Massachusetts, 1998.

[RTJ01] G. Bollella (editor), *The Real-Time Specification for Java*, Addison-Wesley, Reading, Massachusetts, 2000.

[SAR02] T. Saridakis, "A System of Patterns for Fault Tolerance," presented at the 2002 EuroPLoP conference. *http://hillside.net/patterns/EuroPLoP2002/Saridakis.zip.*

[SCH96a] D. Schmidt and C. Cranor, "Half-Sync/Half-Async: An Architectural Pattern for Efficient and Well-Structured Concurrent I/O," in [PLPD96], pp. 437–59.

[SCH96b] R. G. Lavender and D. C. Schmidt, "Active Object: An Object Behavioral Pattern for Concurrent Programming," in [PLDP96], pp. 483–99. Also discussed in [RIS01], pp. 347–52.

[SCH00] D. Schmidt, C. O'Ryan, M. Kircher, I. Pyarali, and F. Buschmann, "Leader/Followers," presented at the 2000 PLoP conference. *http://hillside.net/plop/plop2k/proceedings/ORyan/ORyan.pdf.*

[SEL98] B. Selic, "Recursive Control," in [PLPD98], pp. 147–61.

[SGW94] B. Selic, G. Gullekson, and P. Ward, *Real-Time Object-Oriented Modeling*, John Wiley & Sons, New York, 1994.

[UTAS01] G. Utas, "A Pattern Language of Call Processing," in [RIS01], pp 131–69.

Index

Robust Communications Software G. Utas

ISBN: 0-470-85434-0 (HB)

D

Printed and bound by CPI Group (UK) Ltd, Croydon, CR0 4YY

16/04/2025

14658561-0001